Islamic Theory of Evolution

The Missing Link between Darwin and the Origin of Species

T.O. Shanavas

BRAINBOWPRESS

Library of Congress Number: 2004098291

ISBN: 978-0-9825867-0-9

Cover Design: Uğur Şahin
Grpahics by C.A.D. Graphics, Inc.
Text design and composition by John Reinhardt Book Design

The views expressed in this book solely belong to the author and do not necessarily reflect the views of the publisher, and the publisher hereby disclaims any responsibility for them.

Printed in the United States of America

10 9 8 7 6 5 4 3

CONTENTS

THE QUR'AN 29:20

"Say [O Muhammad]: 'Travel in the earth, then see, how Allah originated creation; and then He produces a later creation.' Allah is omnipotent"

What you read in this book is what I heard from the people who "traveled in the earth to see how Allah originated creation, and then He produces a later creation."

ACKNOWLEDGMENTS

I AM GRATEFUL TO MY WIFE, Faizi, and my children, Zaifi and Reema, for their understanding, and encouragement despite the fact that work on this book took considerable family time away from them. I have borrowed ideas, words, and phrases from many sources and if I have missed to acknowledge of any source, it is not intentional. I am thankful to Professor Samuel Abraham of Siena Heights University in Adrian, Michigan, who provided me with some of the bibliographic materials for the book. I am greatly obliged to the late Imam A. M. Khattab of the Islamic Center of Greater Toledo for his review and critique of the manuscript. Sincere appreciation goes to the Islamic Research Foundation and its president, Professor Ibrahim Syed of the University of Louisville, who played a major part in encouraging me to pursue this work. Without him, this book would never have been completed. Finally, I am grateful to Laurie Rosin and Nada Najjar for their superior editorial services. I greatly appreciate John Reinhardt Book Design for the outstanding service.

FOREWORD

THIS WORK should be used as a textbook in every college, especially in countries with predominantly Muslim populations. It should be translated into all major languages. It provides irresistible evidence and ingenious arguments in educating the reader on historical, philosophical and scientific misconceptions and distortions. If you think that the evolution of the species is mere atheistic theory, this book will challenge and ultimately refute every aspect of your argument. You will be surprised to learn that it is just the opposite: evolution is the divine law in the creation of all living beings.

Dr. Shanavas' book was previously published under the title *Evolution and/or Creation: an Islamic Perspective*. We decided to publish the revised version under a new title, *Islamic Theory of Evolution: The Missing Link between Darwin and the Origin of Species*. The cover picture realistically reflects the common attitude among the proponents of the theory of evolution since they ignore the important contribution of Muslim scientists and abuse the theory by associating it with unwarranted philosophical conclusions such as atheism. Dr. Shanavas respects Charles Darwin as one of the greatest scientists.

In this landmark book, Dr. Shanavas defends the theory of evolution both scientifically and theologically and provides historical background that has been eradicated from public memory. Though a great majority of people, regardless of their religion, consider

Charles Darwin as the originator of the idea of evolution, Shanavas reminds us that Darwin (1809-1882) and his grandfather Erasmus Darwin were influenced by the work of Muslim scientists who lived centuries before them.

Dr. Shanavas addresses a major philosophical issue, the perceived conflict between random events and divine intervention and aptly resolves the problem. Shanavas explains that God limits His omniscience as well as His omnipotence to provide a degree of free will for His creatures. At the same time, God maintains His omniscience by making time relative, and without interfering with the immutable laws of nature, He allows us to enjoy the gift of freedom.

Dr. Shanavas' book is recognized by experts within the field. For example, Professor Allene Phy-Olsen, in a recent book, *Evolution, Creationism, and Intelligent Design*, recommends Shanavas' book as "basic to any discussion of the Abrahamic religions and modern science."

Edip Yuksel

INTRODUCTION

WHEN MY SON, ZAIFI, was in high school, he said, "Dad, you send me to the best school around our home to study science. You send me to the Islamic Center to study the Quran. Science says that human beings evolved from the world of apes, but the Islamic Center teaches us that humans were initially created in heaven and came to earth fully formed. What is the truth?" I could not answer him immediately. His curiosity motivated me to study the concept of Creation in Islam and led to the writing of this book. My search to answer my son led me to conclude that science and religion need not be competing ideologies.

This book has several objectives. First it describes aspects of the Islamic view of Creation and seeks to arouse interest in the integration of modern knowledge with religion. Second, it brings to light important historical facts about the theories of evolution and uniformitarianism (a subject treated in detail in chapter 4) that have been dismissed or forgotten. Although Western scientists claim, "The morality of science is reporting the truth,"[1] they recount the history of evolution starting only from Lamarck and Darwin. They ignore the scientific contributions of Muslim scholars or present them as gifts from the Greeks to humanity. I seek, therefore, to set the historical record straight, for surely an ethical interpretation of scientific evidence requires accurate reporting of the truth as a whole.

1

Third, the book introduces an additional dimension into the evolution-creation debate over the policy of teaching creationism in the schools. In the United States, Christian fundamentalists are pressuring every level of administration of our educational systems, Congress, and the courts to include their version of creationism in the school curriculum. This book serves to inform fundamentalists and lawmakers of other creationist beliefs that exist among the citizens of the United States of America.

Many Muslims, particularly Adnan Oktar, known as Harun Yahya, lavishly promote fundamentalist Christian creationism among Muslims. Harun Yahya, in cooperation with fundamentalist Christians, is spearheading a massive campaign against the theory of evolution. Many Muslim magazines and newspapers in the United States and other parts of the world are helping Yahya poison Muslim minds against science. Ironically, this practice often provokes an unfortunate backlash, generating skepticism about Islam.

The process by which this occurs is insidious. People like Harun Yahya, using quotes from scientific journals but applying them out of context, mislead those Muslims who are scientifically unschooled into believing the fundamentalist Christian doctrine of the so-called scientific creationism.[2] Following the modus operandi of the fundamentalist Christian organization Institute for Creation Research (ICR), Yahya uses pseudoscience to promote his interpretation of the Quran. The references he cites in his book, if read in their entirety, usually accept and defend evolution. But he routinely selects just one sentence from an article, a line that can be construed to support his arguments, and uses it as a scientific reference. Like the ICR, he distorts single news items from popular journals to "prove" his conclusion. He conveniently ignores the rest of the article or other articles in the same issue of the journal that support evolution. Such pseudoscience promulgated by Muslim cohorts of fundamentalist Christians lead young Muslims studying science at American universities into distrust or rejection of Islam as a whole because they soon discover the flaws and half-truths in this imported "scientific" creationism.

Sadly, the young Muslim men and women studying in Western universities get little guidance from the imams, the religious guides of each Muslim community, at their mosques. Whether in the West or the East, most imams have little knowledge of science. When asked about human evolution, they say it is a theory, but they cannot render a scientific definition of the word. Imams in general are incapable of giving informed, credible answers to our budding anatomists, embryologists, paleontologists, molecular biologists, physicists, chemists, and physicians.

Many Muslim parents living in the West have scientific training, which they use in their daily life. But they, too, can be poor guides for inquisitive young Muslim minds. They tell their children that scientific theories may be proved wrong at a later time. They ask these intelligent young men and women to listen to their imams and believe what they say. Simultaneously they not only encourage but also often pressure their children to major in scientific fields. What an amazing contradiction—lauding the study of science but rejecting its discoveries!

I hope that by means of this book Muslims will remind themselves of their contributions to the theory of evolution predating Charles Darwin and stop propagating the Christian fundamentalist doctrine of creationism in their community. I hope also that Muslims will join hands with mainstream scientists to prevent the corruption of public-school science curricula with pseudoscience. And finally, I hope it may in some measure rescue young Muslim science students from the tragic quandary in which they now find themselves.

Imams of madrasas (Arabic for schools)—who turn innocent Muslims into monstrous killers, as we have seen in the September 11 incident—preach that Adam was created ex nihilo in heaven, the eternal Garden of Final Return, without any connection to other creatures of the animal kingdom. This book will expose the dichotomy between the imams' kutuba (sermons in mosques) and simple, literal readings of the Quran regarding Creation. I hope thereby to unveil the ignorance of these imams, so all Muslims might see and understand the true teachings of their faith and so young Muslims will not trust the imams who distort Islam.

I have pursued the study of Creation for both religious and personal purposes. The Quran teaches Muslims that those who are pious shall worship God through the "signs in the Heavens and the earth and in [themselves]." The "signs of God" to which the Quran constantly refers are primarily the phenomena of the natural world:

> **3:190–191** Behold! In the creation of heavens and the earth, and alternations of night and day,—There are indeed Signs for men of understanding,—Men who celebrate the praise of God, standing, sitting, and lying down on their sides, and contemplating the (wonders of) creation in the heavens and the earth: 'Our Lord! Not for naught hast Thou created (all) this! Glory to Thee! Give us salvation from the penalty of fire.'

Finally, I consider my effort to illuminate the Muslim portrait of creation as an act of worship (*ibadath*). The prophet's tradition states: "Allah said: 'I was a hidden treasure; I wanted to be known; therefore I created the world so that I would be known.'"[3] As the tradition suggests, Islam does not clearly delineate between the natural and supernatural. The physical universe is an integral part of the Muslim's religious universe. Humanity has to explore the universe and its laws to know God. Hence, understanding the spiritual significance of an active nature (*kiyan*) is an important stepping stone on the road to Islamic spirituality. The sun, moon, and stars are not mere heavenly bodies but religious realities that participate in the Islamic vision of the universe.[4] In the Quran, God even takes them as witness on the Last Day: "So, I call the receding stars to witness." (81:15) Both the Quran and nature urge human beings to worship God: "We shall show them our portents on the horizon and within themselves until it will be manifested unto them that it is the truth…" (41:53) Therefore, a contemplative attitude toward nature is a form of worship.

The main theme of this work is an inquiry into the creation of humanity and the physical universe. According to scripture, God created other, nonphysical beings as well -angels and jinn. The Quran states that God created angels from light; traditionally they are described as almost robotic, though luminous, servants of Allah with no free will. Jinn, by contrast, are created from fire-smoke, and unlike angels they are endowed with free will. This book does not discuss the creation of supernatural beings, because they are imperceptible and therefore cannot be studied through physical observation.

Even though I have disagreed with many Muslims, Christians, and Jews on some of their interpretations of the story of Creation, I do not claim to be more virtuous than them or other religious people. I share their belief that existence emanated from God, but I do not believe in a sectarian God. If God is as just as all religions claim, God cannot be sectarian. On the Day of Judgment, a just God's reward or penalty to human beings cannot be based on which group they were born into or chose to follow but solely on their righteous deeds and faith. The following verse from the Quran confirms this view:

> **20:15**...For every soul will receive its reward by measure of its endeavor.

In this verse, the phrase "every soul" negates the contemporary Muslim claim of exclusivity for God's love and reward, unless Muslims make a parallel assertion that Christians, Jews, and others do not have souls and Muslims are the only ones with souls.

The Quran elaborates further:

> **4:123–124** It is not your [Muslims'] whims, nor by the whims of the People of the Books [Jews and Christians]. Whoever does misdeeds shall be recompensed for it, and shall not find for him, apart from God, neither patron, nor helper. And whoever, whether male or female, does righteous deeds while believing, shall enter the garden and not be wronged a jolt.

> **2:62** Those who believe (in the Quran), and those who follow the Jewish (scriptures), and the Christians and the Sabians—any who believe in God and the Last Day and work righteousness shall have their reward with their Lord: on them shall be no fear, nor shall they grieve.

> **3:113–115** Yet they are not all alike; some of the People of the Book (Christians and Jews) are a nation upstanding, that recite God's signs in the watches of the night, bowing themselves, believing in God and in the Last Day, bidding to honour and forbidding dishonour, vying one with the other in good works; those are of the righteous. And whatsoever good you do, you shall not be denied the just reward of it; God knows the god-fearing.

The Quran even goes further to counsel the believers:

> **5:82** And you will find those nearer in affection to those who believe (i.e., to the Muslims) who say, 'We are Christians', that is because some of them are priests and monks, and because they are not haughty.

Finally, the Quran describes the Torah and the New Testament as books of guidance and light. It even states that Allah allows the people to practice different religions and to compete with each other in doing good deeds. I have added the underscoring for emphasis:

> **5:44–48** We sent down the Torah which contains guidance and light…Later, in the train (of Prophets), We sent Jesus, son of Mary, confirming the Torah which had been (sent down) before him, and gave him the Gospel containing guidance and light…And to you [O Muhammad] We have revealed the book [Koran] containing truth, confirming the earlier revelations…To each of you [Muhammad] We have given a law, a way and a pattern of life. If God had pleased He could surely have made you one people (professing one faith). But He wished to try you and test you by that which He gave you. So try to excel in good deeds. To Him will you all return in the end, when He will tell you of what you were at variance.

How true these verses ring! I have been acquainted with Jews, Christians, Muslims, Hindus, and people from other religious denominations. Many are kind, compassionate, and willing to sacrifice worldly riches for the benefit of all of humanity. Many are less envious and more righteous than I am. Albert Sabin, the Jew who invented the oral polio vaccine; Mahatma Gandhi, the Hindu and man of peace; and Alexander Fleming, the Christian inventor of penicillin, were great men who committed themselves to the welfare of human beings. Father Damien, who served the lepers; former US president Jimmy Carter, a Christian peacemaker who builds homes for the homeless; and Mother Theresa, a Catholic nun who helped and comforted the poor in the slums of Calcutta, are also fine examples of individuals who dedicated their lives to the service of humanity and worshipped God through selfless service. The Bible inspired Mother Theresa, Jimmy Carter, Father Damien, and millions of Christians to dedicate their life to relieve the pain and suffering of all people. They vindicate the Quranic dictum that guidance and light can be found in the Holy Bible. I may disagree with some of their religious beliefs, but in light of the Quranic verses cited above, it would be absurd for me to imagine that they would not receive more compassionate and merciful treatment from Allah than I would as a Muslim.

Thus, the conclusions I draw in this work about certain contemporary Judeo-Christian-Islamic interpretations of Creation are not an expression of dislike toward these traditions. They constitute a rational, mostly literal, interpretation of the Muslim Holy Book.

I hope that this book will be useful to scientists who are ignorant of the extensive history of Muslim thought on evolution and to non-Muslims who have little knowledge about the Muslim interpretation of Creation during the classical period of Islamic history. Further, I hope this book creates a sense of curiosity in the mind of Muslims who may have lacked the will to question the irrational pronouncements made by some of today's religious scholars.

Finally and most importantly, I wrote this book with the intention of fostering understanding among all peoples of the earth; I fervently hope that it will not be used by anyone to advance the cause of religious intolerance. Otherwise, I will have failed in my endeavor. I believe that human beings are created with pluralistic and multicultural qualities. We are not here to alter God's creation by force or hateful distortions. Rather, we ought to appreciate the beauty and versatility of this world and affirm the principles that enhance harmony and tolerance among the human race. The Quran supports this harmony in chapter 49, verse 13:

> **49:13** O mankind! Lo! We have created you male and female, and have made you nations and tribes that ye may know one another. Lo! the noblest of you, in the sight of Allah, is the best in conduct. Lo! Allah is Knower, Aware.

A NOTE ON TRANSLATIONS OF THE QURAN:

Please note that throughout the text I use literal translations of the Quranic verses, as is common in Muslim scholarship. The translations of the Quranic verses quoted herein should be viewed as the opinion of the translator, however.

Muslims believe that it is humanly impossible to translate the Quran or to reproduce its majestic beauty, indescribable rhythm, and rhetoric. Muslims believe that there is only one Quran, which has been preserved in its original form in its original Arabic language. Thus, any translation or commentary in any other language is considered the opinion of the translator or commentator.

The word Allah is an excellent example by which to illustrate the problems entailed in translating the Quran into English. Allah is the Arabic term for God, the Supreme Creator of all that exists. There is no plural for the word Allah in Arabic. Allah has no partners, sons, or daughters. Allah is self-sufficient, and no deities precede or exceed "him." Even though the male gender is assigned in English translations for Allah, Allah is neither male nor female.

Because it is nearly impossible for an author to extricate himself from his work, my vision of the world has undoubtedly influenced the thoughts expressed in this book. I have applied every effort to remain close to the literal text of the scriptures in drawing an accurate picture of the concept of Creation.

CHAPTER 1

InshaAllah (Allah so Willing) the Islamic Metaphysics of the Future

I N THE LIFE of devout Muslims, a day never passes without using the Arabic phrase *Inshah Allah* (God-So-Willing) at the end of any conversation about future events. Without an understanding of the meaning of this phrase, we cannot begin to comprehend God's relationship with His creatures, the concept of Creation, and the role of free will. Nor can we offer a rational, internally consistent argument against materialist exclusion of God in the evolution of life and the universe.

Materialists among scientists argue that biological evolution is an "inherently mindless purposeless process."[1] They preach that impersonal laws rule the universe and that atoms are at work in the operation of life. Biologist and atheist Richard Dawkins insists that contingency and natural selection, operating over a long period of time, account for evolution. Dawkins assumes that blind forces of physics, chemistry, and natural selection are sufficient to explain the origin and expansion of life.[2&3] He asserts that the unfolding of life serves as the results of selfish desires from genes to increase their opportunities for survival and reproduction. Similar opinions prevail among other practitioners and admirers of science who argue that there is no reason to include God in the evolution of life. One extremist states that "materialism is absolute [and] we cannot allow a Divine Foot in the door."[4]

Such fervency stems from an unshakable, unwavering faith in the Law of Causality, which most people acknowledge and which states that a given cause always produces the same effect. Gravity always pulls an apple down to the earth; spring season melts snow; drought brings destruction of crops. Chemical reactions in any organism, amoeba or human are explainable by the same laws of physics and chemistry that govern the universe.

Based on causality, scientists maintain that the future is predetermined and can be predicted through accurate knowledge of past causes. The laws of nature, they argue, are invariant, and scientific observation reveals the past as the product of those laws. Any natural event that departs from the anticipated effect of a uniform cause is classified as an "accident." However, scientists' predictions based on observation of matter and invariant laws of nature are limited by their own earlier conclusions and experiences.

To gather data, scientists peer into nature, from atoms to stars, amoebas to humankind, fungi to maple trees—any understandable phenomena within our universe. Science has categorized the collected data, defining disciplines such as paleontology, comparative anatomy, biogeography, embryology, molecular genetics, and so on. The materialists' claim that the unfolding of life serves as a "purposeless, mindless process" is based upon inferences from separately catalogued extrapolations of past experiences. John F. Haught, professor of theology at Georgetown University, calls such materialistic metaphysics "metaphysics of the past."[5]

Haught's outstanding treatise God After Darwin helped me to develop a stronger understanding of Islam and the concept of Creation as described in the Quran.[6] I find his mode of thought to be as perceptively Islamic as his belief system, so I shall apply his metaphysics in an Islamic context.

In the Islamic universe, unlike that which is catalogued by science, the past and the present are not the creators of the future, nor are humans or any other creatures, because "...Allah is the creator of everything..." (13:16). Even creations that we claim as our own emanate from God. The Quran states, "And God created you and what you make" (37:96). Allah created everything—computers, airplanes, cars, and even the atom.

A few important questions Muslims might ask are, "If God is the creator of that which humans make, is God not also the creator of our good and bad deeds? Why should there exist rewards for the pious and retribution for the impious if Allah is the source of our actions? If Allah has sealed the heart of disbelievers from receiving divine guidance (2:7), why does God hold us accountable for our actions on the Day of Judgment?" The answer to these questions lies in Islamic metaphysics of the future.

In the Islamic context, time is a conveyor belt carrying messages from God. The experienced past is irretrievable, while the present is only a fleeting moment that we cannot hold. On the other hand, we experience the continuous coming-on of the future. The future in the material world does not exist until it is created. The future is not simply the birth of a moment. *Future* means the yet-to-be-born or created moment packed with contrasting or diametrically opposite possibilities. Each moment presents itself to us simultaneously as different kinds of information. Islamic faith decrees that God is the source of all information. Therefore, Muslims pray, "My Lord, augment me in knowledge." (20:114). Each moment brings each of us hope or fear, success or failure, pain or pleasure, routine or surprise events.

Islamic teachings regarding the coming of future events are grounded in the phrase *Inshah Allah* (God-so-willing) and the verse, "And never say about anything, 'Behold, I shall do this tomorrow,' without [adding] 'if God so wills'" (18:23). Muslims say *Inshah Allah* (God-so-willing) after every statement pertaining to the future, even for simple tasks such as meeting a friend at 4 P.M. tomorrow. Muslims believe that the future is not simply born without cause. It occurs only when and if God creates arriving moments loaded with information. Our planning and our desires may or may not be what God is going to present to us in our future. Allah states: "...and they contrived, and God contrived, but God is the greatest of contrivers." (3:54)

All creatures participate in actualizing the possibilities contained within future moments into visible monuments of divine creation. The present is the pivotal moment between past and future. God tests us by asking us to make moral choices of possibilities—the good, bad, and neutral, the moral and immoral—that are contained in each approaching moment only as information. These choices have no negative or positive charge in the material world until creatures actualize the possibilities into material-world realities. Thus we are necessary participants in the ongoing process of creation. Otherwise, the stream of future moments would remain only as idle information and possibilities. God calls humans to account for our act of choosing from the alternatives within the arriving future moments and also for rejecting information that appears logically correct to each individual.

Each moment is different for each individual, and the lessons of the past and present are tools that can help us make the right choices. God revealed His words to the prophets and provided holy books to both humans and *jinns* (beings who inhabit a different plane of existence). These scriptures explain what is right or wrong, moral or immoral, preferred or not preferred, and rewarding or punishable. Allah narrates stories of the past in the Quran in order to help humans and jinn use lessons from history in conjunction with revelations to make the right choices, as He has graced both of them with relatively higher levels of limited free will compared to other creatures. Living creatures have the freedom to actualize only those possibilities contained as information in each moment of the future arriving from God. To label the above belief, we shall borrow John F. Haught's phrase "metaphysics of the future"[7] and modify it to read "Islamic metaphysics of the future." If a future moment arrives lacking in novel possibilities, humans and other creatures cannot change their present condition, which then becomes stagnant and may remain so for an unlimited period of time. Even when God sends our way the messenger in moments with novel possibilities, we will remain unchanged if we do not accept His offerings and revelations.

Within Islamic metaphysics of the future, the universe is always within God's providence. Therefore, God is the creator of all things and the one who brings the possibilities of each moment through the form of information. Nothing comes into existence without the information about it initially available. Inventors and technicians provide the information they receive to various sectors of society— such as politicians, heads of corporations, and others—guiding them towards the scientific basis for managing their vocations. Those who understand the information can then actualize it into cars, airplanes, nations, and so on.

The metaphor of a factory worker illustrates our relationship to God. While the ordinary assembly-line worker can choose the manufacturing plant in which he wishes to be employed, factory workers have no freedom to manufacture any products of their choice; they must assemble a product using components coming through the conveyor belt of the factory. The physical and spiritual universe is the manufacturing plant of Islam owned by God, the supreme Scientist and Technician. Here, creatures at large and humans in particular are like assembly-line workers. The chain of flowing moments into the coming future is the conveyor belt, which delivers the raw materials (possibilities as information) necessary for the making of many products. In this factory, the worker is free to select any of the components (possibilities) from the conveyor belt (arriving future) and actualize those possibilities into visible monuments of God's creation.

If there is no flow of information from scientists and technicians, the assembly-line worker is unable to produce anything. Even the factory would not exist. Likewise, humans or any other creatures cannot produce or act upon the world until the future moment arrives with possibilities as information from God. Belief and disbelief in God also come as possibilities in the flow of moments from the future. If human beings choose and accept disbelief in God, their minds become unreceptive to divine revelations until they are willing to give up their disbelief. Consequently individuals opt to receive reward or retribution in the Hereafter universe (al-Akhirah) based upon their earthy choices.

The universe from the Big Bang to the Big Crunch is a maze. On both ends there exists a singularity in which all matter is condensed into a mathematical point. The maze is made out of alleys, roads, highways, and byways that lead to different futures for the universe

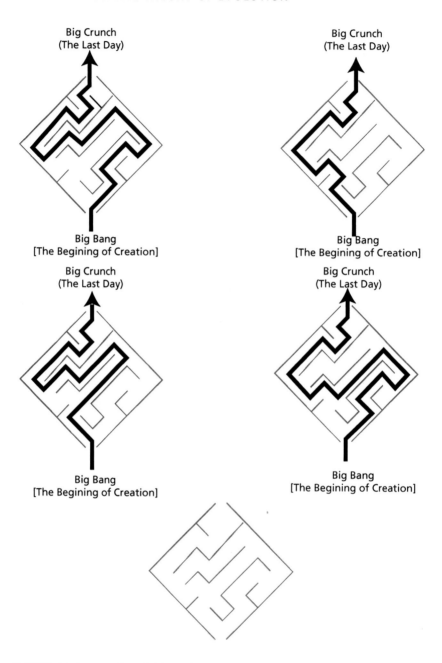

FIGURE 1-1. Maze of evolving Universe. In this maze of evolving universe, there are four potential routes that the universe could travel from Big Bang (The beginning of Creation) to the Big Crunch (the Last Day).

and its components before it faces the Big Crunch (the Last Day). (Figure 1-1). God has already mapped out (created) all the possible and available futures that we can choose, but it is still up to humans and other components of the universe—day by day or moment by moment—to decide for themselves which alleys or roads or highways to step into. God, the Merciful and Benevolent, does not interfere or force us into making choices and voluntarily limits His Absolute Power as stated in the Quranic verse: "And had your Lord willed, whoever in the earth would have believed altogether. Will you then coerce the people to become believers?" (10:99) God knows that free will would remain nonexistent for His creatures without putting limitations on His omnipotence and omniscience. Creatures would not be able to choose when the future arrives with possibilities from God without voluntary self-control of His power and absolute knowledge. Therefore, *Al-Rahman* (The Beneficent) and *Al-Rahim* (The Merciful) set a voluntarily self-imposed limitation on His omniscience and omnipotence to create free will for His creatures. Because the self-imposed limitation is voluntary, it does not imply any inherent limitation in God's ultimate power and omniscience. At the same time we are free to choose and actualize any of the worldly possibilities available to us—atomic power, computer technology, biological engineering—but our future remains limited by possibilities that God has in store for us. In another words, Allah knows all available futures for creatures, but in order to create free will for His creatures, God, being the Most Merciful and Most Benevolent, voluntarily opted not to know which future path that His creatures would choose to step into until it is done.

In mandatory daily prayers, Muslims recite the opening chapter of the Quran, *al-Fatihah*. "Guide us (O Lord) to the path that is straight, the path of those You have blessed, not of those who have earned Your anger, nor of those who have gone astray." These verses imply to many Muslim and non-Muslim minds that God does lead some of us astray. Based upon the Islamic metaphysics of the future, God presents contrasting and diametrically opposite possibilities within arriving moments of future, and He also states in his revealed books which possibilities please Him or draw His anger. We are thus graced with guidance as to how to travel in the righteous path, by choosing righteous possibilities from the arriving moments of future. Similarly, we are offered guidance about which choices from the possibilities in the arriving moments of the future lead us astray.

Even though God is the source for good and sinful possibilities as well as the source of guidance for selection or rejection of possibilities that are respectively righteous or evil, the possibilities remain as idle information until we actualize any of the information from the arriving future into the reality of the material world. God does not interfere in the free choice of His creatures from among the available possibilities within arriving future moments. The sinners' own confessions, as repeatedly mentioned in the Quran, includes statements that their great leaders or the Devil misled them. Not once do they put forward the excuse that God Himself misled them: "And they (shall) say, 'Our Lord, we obeyed our masters and our notables, but they led us astray from the path.'" (33:67) "And those who disbelieved shall say, 'Our Lord, show us those who led us perverse of men and jinn, and we shall put them under our feet that they may be at the bottom.'" (41:29) If God had really led them astray, the sinners' best excuse on the Day of Judgment would have been that they did not deserve to be punished because God Himself led them astray. Therefore, based upon the Islamic metaphysics of the future, the phrases "God misled" or "God guided" refer to the idea that God simultaneously created good and bad possibilities as information within the arriving future moments. Even though God is the source for righteous and evil possibilities, the option to choose one or the other belongs to the creatures.

Based upon the far-reaching meaning of the phrase *Inshah Allah* and the concept of "metaphysics of the future," accidents or contingencies are novelties coming from God, even though these novelties appear as random to a human mind fixed in the materialists' "metaphysics of the past." For example, God created for Hind bint Utba, one of the foremost enemies of the Prophet Muhammad (peace be upon him), the possibility to join the distinguished company of the Prophet as Hamzah did, or the decision to kill and mutilate Hamzah. God did not compel Hind to choose any one of the paths. God created good and bad choices within Hind's arriving future, and she decided to actualize the bad choice to kill and cannibalize him.[8]

The Quran supports the above when it states: "Surely God does not wrong anyone, they wrong themselves." (10:44) No conflict arises, therefore, between the belief that God is the creator of the world and the belief that all creatures are endowed with limited free will to act upon the possibilities that come through the flow of moments of the future. Such a construction of the universe along with merciful voluntary limitation of His power distances God from the concept of a tyrant.

If we take a global snapshot of all earthly creatures, we witness a universe with a mixture of pain, happiness, and peace. Christian theologian and biochemist Arthur Peacocke describes the universe as a musical play wherein actors (God's creatures) in their individual hierarchical ranks freely choose their roles from possible scenarios presented to them by the choreographer (God).[9] Chance and unpredictability are inevitable and are, in fact, built into such an atmosphere. In real life, however, the pain and suffering we experience as a result of the free choices of all creatures favoring their selfish benefits serve as God's way to test, perfecting, and prepare us for the eternal world of absolute happiness and peace. Sufi poet Jalaluddin Rumi describes this universe as a battlefield where atom struggles with atom like faith against infidelity. In this struggle some benefit and others suffer.

Islamic metaphysics of the future blends chance, unpredictability, and limited levels of free will to form our universe, within the providence of a Most Compassionate, Most Just, and Omnipotent God. In this universe, God proposes and humankind and other creatures together dispose. God's rewards or retributions in the Hereafter world are based upon our intention and selection and actualization of the righteous or evil information within the messenger-moments called *the future* into visible monuments of God's creation in the material world. This conclusion is supported by the following verse:

> **4:85** Whoso interveneth in a good cause will have the reward thereof; and whoso interveneth in an evil cause will bear the consequence thereof. Allah overseeth all things.

CHAPTER 2

The Concept of Time: New Physics and the Quran

I ONCE ASKED my seventh-grade teacher, "What is time?" Unfortunately, my question enraged him because he thought I had made a serious grammatical error. He advised me that the correct way to inquire was to ask, "What is the time?" In those days of corporal punishment in schools, I was afraid to explain to him what I meant. Time passed. Days grew into months, months into years, and years into decades, but my curiosity did not change. I knew that time is what a clock measures, but I continued to ponder the question "What is it that the clock measures?" This question and many more became the focus of my attention when I started to study the Quran, the Holy Scripture of the Muslims.

A Muslim cannot ignore nature and its laws. Why not? Sayyed Hossein Nasr, an eminent Islamic scholar, answers that question very eloquently:

> "The Quran depicts nature as being ultimately a theophany which both veils and reveals God. The forms of nature are so many "mosques" which hide various Divine Qualities while also

revealing these same Qualities for those whose inner eye has not become blinded by the concupiscent ego and centripetal tendencies of the passionate soul."[1]

The Quran and the physical universe are "twin manifestations of the divine act of Self-revelation."[2] Viewed as a text, the universe is a "written scroll" (21:104) with information that must be read according to its meaning. The Quran is its counterpart, a text in human vernacular that bids us to explore and coexist with the universe without damaging it. The Quranic verses are called *ayat* (signs, evidences, lessons), as are the phenomena of nature. The earth, sky, mountains, stars, oceans, ships that float upon it, and all the living creatures in this universe are ayat (signs, evidences, lessons). Both the Quran and the phenomena of nature are direct communications from God to mankind.

According to the Quran, the duty of every believer is to use his or her God-given intelligence to understand the universe and its elements, with the aim of reinforcing his or her faith in God and in His power. The Quran draws our attention to the fact that the universe, as a whole, is here for human beings to explore. Through such exploration, humanity can understand the nature of matter, the physical laws that govern it, and ultimately God Himself. The following verse supports this belief:

> **51:20–21** And in the earth are signs [ayat] for those who have firm faith, and in your own selves. Do you not discern?

Moreover, in the ayats quoted in the introduction (3:190-193), the Quran commands humanity to ponder the wonders of creation in the heavens and the earth. The verses warn Muslims about severe retribution for those who shy away from studying the ayats contained in the universe. Those who do not seek rational knowledge through observation, listening, and hearing are depicted as the lowest of beasts in the sight of God.

> **8:22** The worst animals before God are the deaf, the dumb, and those who do not use their reason.

According to the Prophet (peace be upon him), the Quran does not benefit anyone who does not have the basic knowledge to understand it. The Hadith (sayings attributed to the Prophet Muhammad) assert:

> "To rise up at dawn and learn a section of knowledge is better than to pray one hundred *rakah* [repetitions of prayer with movement]; it is better than the world and its contents; knowledge is a treasure-house and its key is inquiry. So inquire, there are rewards for four persons: the inquirer, the learned man, the audience and their lovers; to be present in an assembly with a learned man is better than praying one thousand rakahs. The messenger was asked, "O Messenger of God! Is it better than the reading of the Quran?" He replied, "What benefit can the Quran give except through the knowledge?" [3]

Based on the above verses from the Quran and the Hadith, the religious duty of every Muslim is to understand all creations of God so that he or she can comprehend God. Moreover, we cannot understand and benefit from the Quran with knowledge of the external world or natural world. Time, being one of the created phenomena of the universe, serves among the subjects of contemplation and analysis by Muslims. Here is another reason to study the nature of time. Many Muslims, without ever inquiring into the nature of time, believe in the suddenness by which the universe and mankind were created within our earthly time frame. This belief has led them to the rejection of paleontological, biochemical, and other scientific data that biologists use to infer the evolutionary birth of life, plants, animals, and human beings. When Muslim students of science point out the dichotomy between science and the contemporary Muslim teachings in their mosques, many imams and Muslim parents tell students that reason cannot be trusted and that the human intellect is unreliable—but they forget that they are using their intellect to say so! Why should anybody trust their opinion on any subject when they distrust the capabilities of their mind? Human beings must exercise their intellect to choose the imams' religion over other religions or vice versa. Otherwise, how can one determine who has chosen the true path of God?

Many would argue that the choice of one's spiritual path is best made in the heart, not in the mind, and that our limited mind is no more capable of determining the "true path of God" than of grasping God's eternity or omniscience. Many Muslims argue that our heart can experience this, but not in the form of thought. Here Muslims are forgetting the Quranic verse equating those who do not use their intellect to instinctive animals:

> **7:179** They have hearts [hearts are often used in the Quran as synonymous with minds], but they do not comprehend with them; and they have eyes, but they do not perceive with them; they have ears, but they do not hear with them. They are like cattle; nay, they are even more perverse. Those are they, the neglectful.

Hence, use of intellect and human understanding of the world remains very important, as is further illustrated by these sayings of the Prophet (peace be upon him):

> "God hath not created anything better than reason, or anything more perfect, or beautiful than reason; The benefits which God giveth are on its account; and understanding is by it, and God's wrath is caused by disregard of it.[4]

> "It is not a sixth or a tenth of a man's devotion which is acceptable to God, but only such portions thereof as he offereth with understanding and true devotional spirit; Verily, a man hath performed prayers, fasts, charity, pilgrimage and all other good works; but he will not be rewarded except by the proportion of his understanding."[5]

At the same time God is merciful to his limited creatures and their limited intellect. God rewards those who genuinely search for the truth even though they may have drawn wrong conclusions from their search. This is reflected in the following saying of the Prophet (peace be upon him):

"When a judge gives a decision, having tried his best to decide correctly and is right, there are two rewards for him; and if he gave a judgment after having tried his best to arrive at the correct decision but erred, there is one reward for him.[6]

Similarly the Quran states,

33:5 "And there is no sin for you in the mistakes that ye make unintentionally, but what your hearts purpose (that will be a sin for you). Allah is ever Forgiving, Merciful."

The genuine search for truth requires a will to doubt. Such skepticism is not anti-Islamic but a fundamental teaching in the Quran. When Prophet Abraham (peace be upon him) said to Allah, "My Lord, show me how you raise the dead," God asked, "Have you not believed?" He [Abraham] said, "Yea, but to make my heart well assured." (2:260). The Prophet's (pbuh) commentary to this verse is, "We have more rights to be in doubt than Abraham (pbuh) when he said, 'My Lord! Show me how thou wilt raise the dead.'"[7] This saying of the Prophet (pbuh) and the Quranic verse make the point that even though human intellect is limited, the use of our intellect is as important to the search for the truth as revelation.

Human beings since the beginning of civilization have pondered the concept of time. The unyielding irreversibility of the passage of time is borne in human beings as the certainty of death. Unlike other life forms, we know that our life could end at any moment, and even if we attain all our earthly expectations, our success is inevitably followed by eventual decay and, in due time, death. The notion of time in ancient civilizations was much different from that of today. People did not perceive time as a linear continuum that stretches into a continuous future. They pictured it as cyclical in nature and therefore believed that historical events also followed a cyclical pattern.[8]

The Greek philosophers, including the Orphics, Pythagoreans, and Plato, held the view that people are reborn in the flow of time. They taught that our perception of one lifetime per human being is an illusion due to the loss of memory about past lives upon rebirth. Some Greek philosophers, such as Pythagoras and Empedocles, were said to be able to recollect their previous lives. Buddha also recollected all his previous lives. Hindus believe that people repeat cycles of birth and death until they break the succession through their vigorous ascetic performance. On observation of repetitive earthly phenomena such as rotation of the day and night or the four seasons, the ancients inferred that time and everything else including human birth and death are cyclic.[9] Pre-Islamic Arab pagans considered time as a deity (ad-dahr) that exists from eternity to eternity and dispenses good and ill-fortune to mankind (45:24).

The monotheistic religions—Judaism, Christianity, and Islam—led the way to the concept of linear time. The chronology in the Jewish and Christian scriptures implies that the universe was created in 4,004 b.c.[10] The Muslim scripture does not reveal a specific time for its creation, even though some Westerners hold that the Muslims also believe in a young universe. For example, Professor Stephen W. Hawking, one of the greatest minds of the twentieth century—hailed as an "equal of Einstein" by Time magazine—states, "According to a number of early cosmologies and the Jewish/Christian/Muslim tradition, the universe started at a finite and not very distant time in the past."[11] Hawking accurately reads the three monotheistic religions when he declares that the universe was created at a finite time, but he errs in assuming that the Muslim estimation of the age of the universe is the same as that held by the Jewish and Christian traditions. Muslims believe that the universe was created in a finite distant past, but they also believe that it will be destroyed sometime in the future. Chapter 3 will discuss this in greater detail. The agreement among the three monotheistic religions on the concept of resurrection—whereby humans live from eternity to eternity—is based on a linear rather than a cyclical chronology.

The invention of the clock in the thirteenth century reinforced the notion of linear time. Since then time has been broken down into units of hours, minutes, and seconds and conceived of as something that progresses from the past to the future. Measurement of time affected the lives of people of all faiths. For example, during the fourteenth century, workers in the cities setup specific times for the beginning and the close of a workday.[12] Isaac Newton believed that he knew what time was. At the beginning of his Principia Mathematica, he wrote: "Absolute, true and mathematical time, of itself, and from its nature, flows equably without relation to anything external."[13] This was the dominant belief among scientists until the twentieth century. Today we know that Newton was mistaken in several respects. Time is not absolute or universal but relative. As physicist Paul Davies observes:

> "Einstein demonstrated that time is in fact elastic and can be stretched and strung by motion. Each observer carries around his own personal scale of time which does not generally agree with anybody else's. Our individual perception of time does not appear distorted to us, but for observers who move in different time frames than ours, we seem to be out of step with their time.[14]

The time between two events at two different locations remains greater for the earthbound observer than for a space-traveling observer. This is called time dilation. The Theory of Relativity states that depending on his velocity, one day for a space traveler can equal a few years or more for an earthbound person. This startling prediction of relativity can be illustrated by the following science fiction anecdote.

A twenty-year-old astronaut takes a trip to a faraway star in a spacecraft that can fly close to the speed of light. His twin brother remains on the earth. After fifty years on earth, the earthbound twin goes to the spaceport to receive his astronaut brother. Both brothers are amazed and startled. The earthbound brother has aged fifty years. He now has gray hair and wrinkled skin. But the astronaut brother has aged only one year. The clock, the calendar, and the biological aging process on the spacecraft slowed down to one-fiftieth of its normal speed. The twins agree that the adventurous space traveler is now forty-nine years younger than his earthbound twin is![15]

According to late Carl Sagan, an eminent astrophysicist, we humans do not experience time dilation in our everyday life. However, nuclear particles experience it when they travel close to the speed of light. Time dilation is measured by their built-in-clocks termed "decay time," and science has validated it with experimental data.[16] The US Naval Observatory in Washington D.C. is responsible for maintaining our time standard. In 1972 a physicist tested the concept of time dilation by carrying along four cesium-beam atomic clocks during his round-the-world trip on a scheduled airline. These clocks could be trusted to a few billionths of a second over the time span of such a voyage. When compared with matching clocks that remained in the observatory, the clocks lost time. The lost time agreed exactly with Einstein's predictions.[17]

Until a few years ago, scientists thought that protons and electrons were indivisible "elementary particles," but when the particles collided at high speeds, scientists discovered that the particles were made up of yet smaller units. In particle-physics laboratories, the time interval for short-living particles like muons to decay into electrons and neutrinos can be recorded. Experiments have shown that the faster muons take longer to die than the slow-moving ones. This is precisely what the special theory of relativity states. Based on such experiments, the twin paradox will not be a paradox at all if and when we are able to travel close to the speed of light. However, the theory of relativity also suggests that nothing can travel faster than light.

According to the general theory of relativity, in a field of very high gravity, time will grind to a halt. Such high gravity exists in black holes and in their peripheries. What is a black hole? When a dying star implodes or shrinks into itself under the relentless and unforgiving influence of gravity, the star becomes smaller and smaller. Indeed, the entire star is crushed out of existence at a single point. Physicists refer to the point of extinction as a singularity (Figure 2-1). At this point, the density of matter becomes infinite and space-time is reduced to a mathematical point. During the shrinking process, the gravity around the imploding star becomes so strong that even the rays of light from the star cannot escape. Because we can see a star only when its rays reach our eyes, the imploding star literally disappears from our vision as it vanishes from the universe. What is left is called a black hole (Figure 2-2).

The periphery of the black hole is called the event horizon. We have no way of knowing what occurs inside the event horizon.[18] Theoretically, time would be frozen for a person in an extremely high gravitational field such as the event horizon while events continue

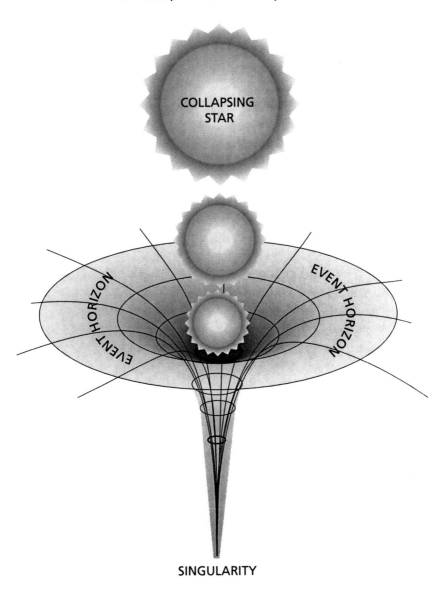

SINGULARITY

FIGURE 2-1. BLACK HOLE AND SINGULARITY. As the star implodes it increases in size. The matter within it is compressed into zero volume, which then becomes a point where nothing exists. Here the curvature of space-time becomes infinite. At this point there is no time (no-when) and no space (no-where). At this point (called Singularity) all theories break down. The black hole, therefore, consists of an unobserved singularity within the event of a star

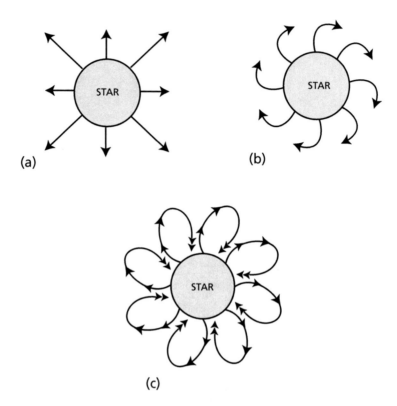

FIGURE 2-2. THE EFFECT OF GRAVITY ON LIGHT RAYS FROM A COLLAPSING STAR. As the intensity of gravity outside the collapsing star increases, the implosion of the star accelerates. The light rays leaving the star bend toward it. View A depicts the star before it collapse. View B. depicts the bending of light rays toward the collapsing star. View C depicts the event horizon encompassing all the rays of light bending toward the star. For an observer located outside the event horizon, the star looks like a black hole because its light rays can no longer reach the sight of the observer.

to unfold in time for the observer on the earth. The gravitational freezing of time could lead to interesting effects. Suppose, for example, that one of the astronauts started to sing the US national anthem while he was entering the event horizon. The general theory of relativity predicts that the situation for observers on earth (and anywhere outside the event horizon) would appear as though the astronaut has been singing the first note of "The Star Spangled Banner" even after the passage of billions of years. This prediction is flawed, however, because the time dilation would have slowed the sound vibrations to a degree whereby observers outside the event horizon would no longer be able to hear the note.[19] While time remains frozen for the astronaut, events would continue on earth and outside it. The astronaut would not be able to leave the event horizon, because from his perspective history has already advanced outside his space and time. He would literally be coming out before he went in. Similarly, according to the theory, two spatially separated events may appear to occur differently for observers traveling at different speeds. One person may say Event A happened before Event B, and another would say Event B happened before Event A, while a third person might argue that Events A and B happened simultaneously.

When the Quran was revealed fourteen hundred years ago, people did not measure time as we do today. They did not wear watches that divided the day into units of time, because clocks had not yet been invented. Their lives were not regulated by hours and minutes in abstract time but by the natural rhythms of nature, such as changes in the seasons and the position of the sun, even though water clocks, sundials, other devices were available to them. A day was the shortest span of time that had meaning and importance in the lives of the early Muslims, although the length of a day varied. Work began at sunrise and ended at sunset. Before the invention of the modern clock, they divided the day into five variable periods to mark the time for mandatory prayers based upon the position of the sun.

The study of the Quran guides us to understand the concept of time among early Muslims. Such a study is essential because time is inseparable from human activity and thoughts. Muslims believe that God is timeless. For example, late Muhammad Asad, a contemporary Muslim scholar, states clearly: "What men conceive as time has no meaning with respect to God because He is timeless, without beginning and without end, so that 'in relation to Him A day and a thousand years are alike'."[20] This Muslim belief is based upon the verses: "This is God, your Lord; there is no god but He, the creator of all things. So pay homage to Him for He takes care of everything." (6:102) "He begets not, nor has He been begotten." (112:3) "Originator of the heavens and the earth ..." (2:117). Time as a separately identifiable entity within the framework of the universe came into existence with the beginning of the universe. If God is the only entity that has no beginning, God exists without a clock that divides time into past, present, and future. Muslims posit that God sees, hears, and knows all things in perpetual Now. This conclusion is supported by the Quranic verse: "Verily a Day in the sight of thy Lord is like a thousand years of your Reckoning."(22: 47) Therefore, the Quranic description of time and the Muslim concept of time are that time is relative.

The human understanding of time in the Hereafter Universe (*Al-Akhirah*) also suggests a relative nature for time. For example, the following Quranic verses depict the paradox of the twins to which we referred earlier. The astronaut twin is comparable to a person who is resurrected in the Hereafter Universe (*Al-Akhirah*). He feels that his few years of life and the billions of years of lifeless existence between his death and his resurrection have shrunk to a brief moment:

> **79:46** The day these people see seems to them as though they had stayed (in the world or in the state of death) only for an afternoon of a day or its forenoon.

> **10:45** On the Day when He shall gather them (unto Himself, it will seem to them) as if they had not tarried [on earth] longer than an hour of a day, knowing one another; [and] lost indeed will be they who (in their lifetime) considered it a lie that they were destined to meet God, and [thus] failed to find the right way.

> **17:52** "On a Day when He will call you, and you will answer by praising Him, thinking all the while that you tarried (on earth) but a little while."

When human beings are resurrected, they will be asked about the length of their lives on earth, and they will reply that they lived on it for a day or a part of a day. Furthermore, they will attempt to sidestep the question by stating, "Ask those who are capable of counting time." This evasion is an indication of the dissolution of humankind's earthbound concept of time upon resurrection. The following verse suggests such dissolution:

> **23:112–113** Then Allah will inquire from them, "for how many years did you live on earth?" They will say, "We stayed there for a day or part of the day; but ask those who (are able to) count (time)."

According to the verses above, the resurrected human being, like the astronaut twin, will feel that the whole world—its centuries, its ages, its epics, and its events—has shrunk to but a moment, the length of which is known to God only. Human beings will exist in a timelessness that words cannot describe. How can one express time in a place where there is no space-time relationship? Discussing timelessness is like defining time as an entity that flows. If it moves, it must have speed that can be measured. How can we measure it? If we can measure it, how do we express it? Speed, when measured, is expressed as "miles/hour." To express the speed of time, we would have to write "hours/hour," which is, of course, meaningless.

We have seen earlier that according to the theory of relativity, each observer lives within his own personal timeframe which does not generally coincide with anybody else's. Therefore, the Quranic verses do not violate God's (or Einstein's) law of relativity or the physical fact of time dilation. In summary, Quran teaches that time is relative based upon the state of the observer and whoever observes it.

Why did God create time with the material universe? Why did He make time relative? In the construction of a rational universe for humankind, God knew the logical problems worldly existence would cause for His intelligent creature, the human being. Perhaps the major logical problem is the appearance of a contradiction between attributes such as God's Omnipotence and Omniscience, and His Will to create humans with the faculty to Discriminate and choose either goodness or evil. For example, if God chooses to alter the future course of an event, do not His omniscience and omnipotence stand in the way of human free will?

Omniscience is a characteristic of God according to many verses in the Quran. For example, chapter 2:255 states:

> **2:255** God, there is no God but He, the Living, the Eternal Sustainer. Neither slumber takes Him, nor sleep. His is what is in the heavens and what is in the earth. He knows what lies before them and what is behind them. And they grasp naught His knowledge, but of what He wills. His Throne embraces the heavens and the earth and it tires Him not to uphold them both. He is the Sublime, the Grand.

The verses like this and many more suggest that God knows all things at all times. Hence, some atheists and logicians may argue that God's knowledge, being absolute, cannot go wrong. He knew what Timothy McVeigh was going to do even before he was created. In short, they posit that God "programmed" Timothy McVeigh to kill innocent Americans in Oklahoma City. They argue that McVeigh and other evildoers cannot be considered guilty of any sin because they are only preprogrammed machines.

Because an All-knowing God anticipated these logical problems for humans, He created time as a factor in our universe. He made time relative for all earthly creatures, including humankind, but all events in all time frames from the beginning of the Creation to the end of the current universe are current events as far as God is concerned.

Jewish physicist-theologian Gerald L. Schroeder clarifies the above scenario using the true story of a star that exploded, producing a sudden burst of light 170,000 light years from the earth.[21] Astronomers call such a sudden burst of light from an exploding star a supernova. The world came to know about the supernova when a Chilean astronomical laboratory recorded the arrival of the light on earth after 170,000 light years of travel.

Between the time of the supernova and the arrival of the light from it, a multitude of events occurred. Neanderthals came and went, modern Homo Sapiens took control of the earth, civilizations in the Indus valley as well as the Mayans, Grecians, Romans and others peaked and vanished, the Quran was revealed, Muslims developed the scientific method and trained European scientists which ended the Dark Ages and engendered the Renaissance, the Muslim empire crumbled, the United States of America took the leadership of the world—all before the arrival of the light from the supernova on earth.

Schroeder describes an imaginary super-consciousness without mass that travels with the light from the supernova. Let us assume also that our super-consciousness carries an internal clock with no mass as its companion. How much time would this super-consciousness have experienced? How many ticks would the super-consciousness's clock have made? Schroeder answers: "No time would have passed. Not a few years, not a few hours, or a few seconds. Zero time."[22] In this scenario, there is no lapse of time for the super-consciousness, in spite of 170,000 years of recorded history. A similar situation of Perpetual Now or Present exists in an environment with extremely high gravity, as in the case of a black hole. The 170,000 years of the earth, its epics, individual events, Timothy McVeigh and his atrocious bombing of the Alfred P. Murrah Federal Building, and the life histories of myself and yourself are experienced simultaneously by the super-consciousness.

The super-consciousness would definitely reject our claim that there is a history as well as a past and future series of events on earth. He would say that he sees everything that the primitive twenty-first-century human described as the past and the future. All happened in no time in his frame of reference. There is no post- or pre-knowledge for the super-consciousness, only omniscience, because all knowledge occurred at the same time. Physicist Gerald L. Schroeder freely admits, "I don't pretend to understand how tomorrow and next year can exist simultaneously with today and yesterday. But at the speed of the light they actually rigorously do. Time does not pass."[23]

If gravity, speed, black holes, and other material factors can affect and freeze time, is it conceivable for human beings to perceive the timeless world of God, who created the universe and time from nothing with His Enormous Infinite Power? A reasonable person would not deny the absence of time in the presence of God, the source of all forces of the material world. Accordingly, Islamic theology describes a perpetual Now that pervades in the spiritual domain of God, a state of being/consciousness that is more accurate than the "now" of our super-consciousness. God experiences all events of the material world in a perpetual "Now" without the linear flow of time that pervades the material world. God eternally knows all events as well as all their possible directions, deviations, and modifications occurring in the time frame of the material world. Until we fully understand physics' mathematical mystery of a concurrent yesterday, today, tomorrow, and next year, existing together at the speed of light as described by Schroeder, the coexistence of human free will and omniscience of God will remain incomprehensible because human-kind has no experience of timeless existence.

According to the theory of relativity, nothing can travel faster than light; the theory negates any physical communication with a speed faster than light. This limitation in communication between two points is irrelevant from the Quranic point of view. The Quran states: "To God belong the East and the West. Wherever you turn the glory [face] of God is everywhere: All-pervading is He and All-knowing (2:115)." God, being "All-pervading," is present at all points and at all events. He is present in every frame of reference, whether on the earth, in the light from a supernova, in black holes, or in the time frame of the astronaut of the twin paradox. God extends His presence even to our subconscious mind, as made clear in the verse: "Whether you loudly avow the utterance (or not), surely He knows the secret and what is concealed behind that."(20:7)

God's pervading presence in all frames of reference implies instant communication at any particular point we can imagine. When God comes though a particular frame of reference—for example, the earth—human beings picture that He is communicating in our local time. As we have seen in the previous chapter, God allows the next tick of the local clock to occur in order to open the door to the future with possibilities for His creations. Each tick of the clock is a knock at the door of the material world by messenger-moments of the future arriving with God's proposals. His creatures actualize into their timeline these possibilities arriving from the timeless realm of God. Hence, there is free choice for you and me and Timothy McVeigh.

An analogy from the science of molecular biology may better illustrate this concept. Chromosomes, with their component parts called genes, are the seat of all knowledge or information that creates the outward appearance of all living things as well as all chemical processes that take place within them. The genes are made of DNA.

The genetic information in the gene, or DNA, has to be expressed in enzymes (a particular class of proteins) that guide all chemical processes in the cells. Intermediate molecules, called messenger RNA (mRNA), convey the information held by genes to these proteins.

A chromosome's particular knowledge or information is expressed though a chain of events. A gene for specific information is copied into a messenger RNA and sent to a ribosome. The ribosome manufactures (synthesizes) the protein to express the genetic information (or in metaphysical terms, the will) into body parts during fetal development and into the metabolism in an animal or human body. (Figure 2-3).

The knowledge or information within the messenger RNA is a duplicated copy of the knowledge of a particular gene or genes. Messenger RNA does not create new information or knowledge, but it acts as carrier of information. Similarly, God, who knows all possible alternatives available in any given situation, presents copies of His knowledge to His creatures as possibilities packed in each messenger-moments of the arriving future from Him.

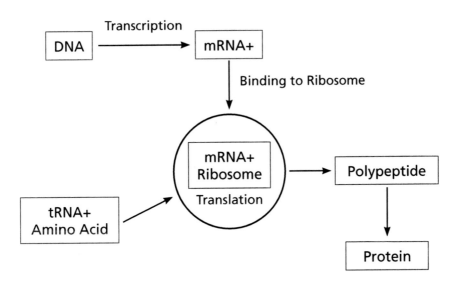

Figure 2-3. GENE EXPRESSION AND PRODUCTION OF PROTEIN.

The ribosome uses the messenger RNA, a copy of a gene or genes, to create proteins, body parts, and external characteristics of all organisms. Human and other creatures also use information presented to them by divine activity in His timeless realm. Creatures actualized a particular possibility from the many possibilities coming as information within each moment of the arriving future, into the monuments of divine creation in the time and space of material world. Each moment of the arriving future is a messenger, like the messenger RNA of molecular biology. While one messenger RNA carries only one possibility, each messenger-moment of the arriving future carries more than one possibility. Each moment can carry contrasting as well as diametrically opposite possibilities or information.

We have seen that Muslims believe that the future never arrives until God creates it. The determining factor in the integration of the messenger RNA into a ribosome is the ribosome's prerogative, which science describes as a chemical property. In much the same way, the determining factor for actualization of any one of the possibilities into a given space and time is the free will that is a divine generosity to God's creatures. God has the full freedom to determine the general direction and trend of the universe, because He creates the messenger-moments of the arriving future and He alone determines the nature of the possibilities to be packed in the messenger-moments for His creatures to actualize into monuments of God's creation. However, omnipotent God voluntarily and gracefully limits His absolute power and omniscience to create free will for His creatures and to objectify their choice of information in the arriving messenger moments from God.

To summarize, two contradictory realities exist:

1. The past, the present, and the future are real in the world of humankind;
2. No past, present, and future exist in the spiritual domain.

There is only a perpetual Now in the world of God. In the world of God, the Big Bang to the Day of Judgment and beyond are current events, and such human linguistic phrases as "what is intended to do" or "going to do" do not exist. Such a state of perpetual "Now" has no future or past. Einstein's own words are more revealing: "[T] he distinction between past, present, and future is only an illusion, however persistent."[24]

Jalaluddin Rumi, a well-respected Muslim Sufi, stated centuries before Einstein in his *Masnavi:*

> "In the space-less realm of light of God, the past, present, and future do not exist. Past and future are two things only in relation to you; in reality they are one. Thy thought is about the past, and future; when it gets rid of these two, the difficulty will be solved."[25]

The poet Rumi rejected the idea of the labeling of time into past, present, and future and described it as only an illusion. Persians describe Rumi's thirteenth-century *Masnavi* as "the Koran in Persian."[26] He may have based his conclusions on the Quranic verses quoted above, or on the verse "He knoweth what (appeareth to His creatures as) before or after or behind them," (2:255) which implies that our past, present, and future are embodied in God's Now. The verse also suggests that God can see, in a moment, numerous events that are spatially separate, although for human beings the same events seem to occur one after another over a long period of time. God sees or knows the past, the present, and the future of humankind as a human being would see what is passing before his eyes at a given moment of "Now." There is no pre- or post-knowledge.

The Quranic verses, Rumi's *Masnavi*, and the theory of relativity inform Muslims that the twenty-billion-year-old universe could be seen as a six-day-old universe as indicated in the Quran. The "biological evolution of life over a period of three and a half billion years" could be experienced in "a blink of an eye" in a different timeframe or in no time. This is theologically supported further by the following verses of the Quran:

> **54:49–50** Indeed everything We have created in measure. And Our behest is one, as the wink of an eye.

CHAPTER 3

Creation and the Fate of the Universe

Cosmology, in the broadest sense, is that branch of learning that conceives of the universe as an ordered system. The term is derived from two Greek words, *kosmos* (order, harmony, the world) and *logos* (word, discourse). For a long time cosmologists thought that the earth was the center of the universe and that the Sun, the Moon, and the planets revolved around Earth. In the sixth century, the Greek Milesean school of pre-Socratic philosophers which included Thales, Anaximander, and Aneximenes taught that the formation of the world occurred as a natural rather than supernatural sequence of events. They were followed by the Atomist school led by scholars such as Leucippus and Democritus who described a boundless universe in which the interplay of atoms brings into existence endless worlds in various stages of development and decay. The Atomists did not admit the existence of a final cause.[1]

Aristotle viewed the universe as eternal, everlastingly undergoing change. He attributed changes on the earth to the motion of the heavenly bodies. Aristotle believed in a prime mover of the universe that keeps everything in motion and in constant change. He saw this being as nonmaterial and without potentiality. Further, he saw the mover of the eternal universe as a perfect entity.[2] I recognize the importance of the Judeo-Christian interpretation of the Creation, because creation-evolution arguments currently are often centered on the Biblical account of the genesis of the universe. I freely admit, however, that my knowledge of the Holy Bible is limited and that many may find the following description of the Judeo-Christian view insufficient considering the number of references in this work.

Until the middle of the nineteenth century, the relationship between science and religion in the West was "largely harmonious, and science took a subordinate role." Western scientists believed that their rightful task was to reinforce the central doctrines of Protestant Christianity.[3] Therefore, the Western cosmology of the nineteenth century was based largely on the Book of Genesis.

In chapter 1, verses 1-31 of the Book of Genesis, we are told that God created the universe in chronological order as follows: Initially the universe was without form. On the first day, God created light and separated it from darkness. On the second day, He created the skies and the oceans. On the third day, He divided the earth into dry land and water and then He created vegetation on the earth. On the fourth day, God created the Moon, the Sun, and the stars. On the fifth day, He created aquatic animals and birds. On the six day, God created earth creatures such as reptiles, cattle, and wild animals, and He also created Adam and Eve. Finally, God rested on the seventh day. The order of events for the creation of the universe is not consistent. Genesis 2 tells us the universe was created in one day. Chapter 1 says God created the fruit trees, the animals, Adam, and then Eve. Chapter 2 says God created Adam, the trees, the animals, and then Eve.

Modern scientific theories do not support the Biblical cosmology of the Book of Genesis, which holds that the Sun and other stars came into existence after the formation of the earth. In contrast to science, the Bible also contends that vegetation and water appeared on Earth before the formation of the sun and other stars, and that birds existed before reptiles. As a result of these conflicting claims, new interpretations of the Book of Genesis started to appear in the West.

Since Christians and Jews believe that God created the universe, they are considered creationists. However, they have no monolithic agreement among them regarding the different interpretations of Biblical verses. Based on their interpretations and attitudes toward scientific theories, Christian creationists fall into four major groups: revelatory theorists, poetic theorists, old-earth theorists, and young-earth theorists.

Those who follow the revelatory theory propose that Genesis 1:11 is not a straight narrative account, but instead Moses' vision of the divine revelation (Moses being traditionally believed as the human author of the Book of Genesis).[4] The second group of creationists follow the poetic theory. They claim that the story of Creation is a deliberately poetic account rendered by Moses, and it therefore should not be understood literally.[5]

The third group is the old-earth creationists, who attempt to justify a move away from literal interpretation. They see humanity as God's special and direct creation. They are distinguished from mainstream Christian accommodation to scientific consensus by their rejection of all or part of the theory of evolution. In general, they accept the idea of an antique earth based on a radiometric method and have different explanations for the age of the earth and universe.

The old-earth creationist groups can be divided into three subgroups. The first, those who follow the gap theory, say that verse 1 of Genesis ("In the beginning God created the heaven and earth") suggests that God created an ordered universe. On the other hand, verse 2 ("And the earth was without form, and void...") reveals the act of giving form and order to an already existing chaotic universe. They believe that in the gap between what was revealed in verse 1 and verse 2, many cycles of creation and destruction occurred over millions of years, which allowed the formation of fossils from the creatures of the destroyed pre-Adamic world.

The second subgroup, who follow the day-age theory, take a metaphorical approach and propose that the English translation of the Hebrew word *yom* as "day" is misleading. They believe that "yom" in Genesis does not constitute a twenty-four-hour earthly day but a longer period of time. However, they reject the evolution of life and human beings. The third subgroup follows the progressive theory and moves further toward accepting the scientific consensus. They see the days of creation as overlapping and of varying length. They concede that the order of events for the process of creation as revealed in Genesis differs from the order apparent in the fossil and geological records. In Genesis, for instance, flowering plants are created before land animals, but they are relative latecomers in the fossil record. Another Biblical inconsistency that progressive creationists accept is the creation of plants on the day before the Sun. Some interpret Noah's flood as a regional rather than a worldwide deluge. Nevertheless, for them, accepted scientific theory cannot explain major events such as the appearance of mammals or other groups. Instead, these events require direct divine intervention. It follows, then, that creation did not happen all at once but over time and that God gradually created new forms of life that eventually culminated in the creation of humans. Some individuals and groups espouse various combinations or variations of the three subgroups. Jehovah's Witnesses, for example, accept both the day-age and gap theories.[6]

The fourth and dominant group propounds the young-earth theory. They follow a highly literal and straightforward reading of the first eleven chapters of Genesis. They believe that the creation of the universe and life herein occurred in a span of six twenty-four-hour earthly days. Some of them believe that creation took place about 6,000 years ago as Archbishop Ussher (d.1656) and other Christian scholars deduced from the chronology of the Old Testament. Other young-earth creationists acknowledge an age of ten thousand or even twenty thousand years for the universe. They believe that God created the universe in its present form and shape, and with all things in it, from nothing. Strict creationists assertively reject any large-scale evolution such as that of humans evolving from apes. They explain the geological record of sedimentary rocks and fossils as remnants of the creatures destroyed during Noah's flood.[7]

Additionally, there are those who believe that God created the universe but relinquished His spiritual authority over the natural and physical sciences. Finally, there are those who take the position that religion and scientific data are two different kinds of knowledge. Evolutionary theory and modern cosmology, they say, are not moral or religious doctrines, and the Bible is not a science textbook but a source of moral and spiritual truth. They assert that Genesis does not give a scientific account of the origin of the universe, but rather conveys how God's creative act began and sustains it.

The Christian theologian Marcus Dods expresses such views in *An Exposition of the Bible*. Dods distinguishes between knowledge regarding the physical formation of the earth and knowledge concerning the relationship between the world and God. The first can be studied through scientific means, the second solely through Scripture (the Bible). He points out that although the details of the story of Creation in Genesis are irreconcilable with scientific discovery, they are "harmonious in their leading ideas." He contends that the compiler of Genesis was less concerned with the process of creation or scientific accuracy and more interested in conveying "certain ideas regarding man's spiritual history and his connection with God."

Dods asserts the futility of trying to reconcile science and religion. He points out that the twofold character of the story of Creation presented in Genesis illustrates two points: that the Bible is irreconcilable with scientific discovery and that it is in conflict with any pre-scientific cosmogonies. He adds, "Science knows nothing about the creation of the sun, the moon, and stars subsequent to the creation of the Earth. Likewise, science rejects the existence of fruit-bearing trees prior to the existence of the sun." Dods confirms that the writer of the Book of Genesis uses his knowledge for the purpose of conveying "his faith in the unity, love, and wisdom of God the Creator." He states, "It is mischievous to build the teaching of the Book of Genesis around discoveries of science because it would not convince independent inquirers."[8]

Big Bang Theory

In the past, people believed that the stars were merely luminous objects affixed to the solid sphere of the sky. In 1718, when Edmund Halley noted that three of the brightest stars had changed their positions relative to the remaining stars, Western scientists realized that stars are moving objects.[9] They began to explore the universe and its origin but even today agree that our knowledge of the origin and evolution of the universe as a whole and its subunits is still very limited.[10] In 1927, George Lemaitre claimed that "the universe has evolved from a hot dense state (about 10,000 billion degrees) and that the evolution commenced with a primordial explosion known as the Big Bang."[11] While the reason for the explosion remains still a mystery, it is clear that all matter and energy which constitute our universe today was compressed into a fiery mass of unimaginable density to form a cosmic embryo.[12]

Physicists and astronomers believe that insight into the structure of the atom will provide the ultimate answer to inquiries regarding the nature of the cosmic egg and the origin of the universe. Atoms have a longer history than that of the earth because they are the elements from which the earth is made. Before atoms came together to form the sun and its planets, they were drifting in the interstellar clouds. Therefore, we can come to understand the genesis of the universe by probing into the constitution of atomic and subatomic particles, and the more we learn about the past of our universe, the closer we get to discovering the secret of its origin. Because it takes thousands or billions of years for light to reach us, events or stars that we see in the sky are hundreds or billions of years old. These stars and galaxies are fossils that guide us to understand the creation and evolution of the universe. Even common objects such as a pebble or a leaf provide us with clues, because they too are made out of galaxies of atoms.

The structure of the atom can be pictured as a tiny solar system in which the sun is the nucleus and the planets are negatively charged electrons, each rotating in a specified orbit (Figure 3-1). The nucleus lies within the confines of the orbiting electrons, and it is assembled out of neutral particles (*neutrons*) and positively charged particles (*protons*). When an electron jumps into another orbit, the photon carries a quantum of electromagnetic force between the orbits. The elementary particles of an atom are made out of older particles known as *quarks*. We can learn more about subatomic particles by accelerating and colliding them in an electromagnetic field, which creates a wide variety of these particles.[13]

Subatomic particles interact with the aid of four fundamental forces: the gravitational force (gravity), electromagnetism, the strong force, and the weak force. Physicists believe that these four forces, in conjunction with atomic particles, serve as the screenwriters of every event in the universe.[14 & 15]

Gravitational force keeps the physical components of the universe together by mutual attraction. A gravitational particle known as a *graviton* carries force between two material particles. Since the graviton has no mass of its own, the force it carries is long-range. The gravitational force between stars and planets is attributed to the exchange of gravitons between the particles that make up these bodies.

Particles called *photons* carry the electromagnetic force by which electrically charged particles interact. Electric charges are of two kinds—positive and negative. The force between two charges is attractive only when one is positive and the other is negative. The electromagnetic attraction between negatively charged electrons and positively charged protons in the nucleus causes the electrons to circle the nucleus of the atom.

When an electron in an atom jumps from a higher-energy outer orbit to a lower-energy orbit closer to the nucleus, the energy difference escapes in the form of a photon. We see the escaped energy as light. The atom cannot exist without the strong nuclear force, which bonds the quarks to form protons and neutrons and holds protons and neutrons together in the nucleus. This force is carried by another particle called a *gluon*, which interacts only with other gluons or with quarks. Finally, the weak nuclear force, carried by a particle known as a *boson*, helps the neutrons inside the nucleus to decay. The strong and weak forces operate only in the nuclei of atoms, but the range for electromagnetism and gravitation is infinite.[16]

Physicists believe that when the Big Bang occurred, all four forces—gravitational, electromagnetic, strong, and weak—united under extremely high temperatures. This process remains a mystery because no laboratory on earth can produce such temperatures. Only the high ambient temperatures of the genesis were hot enough to create such unity. The universe then started to expand and cool. Gravitation became a distinct force 10^{-43} seconds later. Further cooling to 1028° Kelvin, at 10^{-35} seconds after the Big Bang, the strong force became distinct. After about 10^{-10} seconds, electromagnetism and the weak force emerged (Figure 3-2). When the universe was

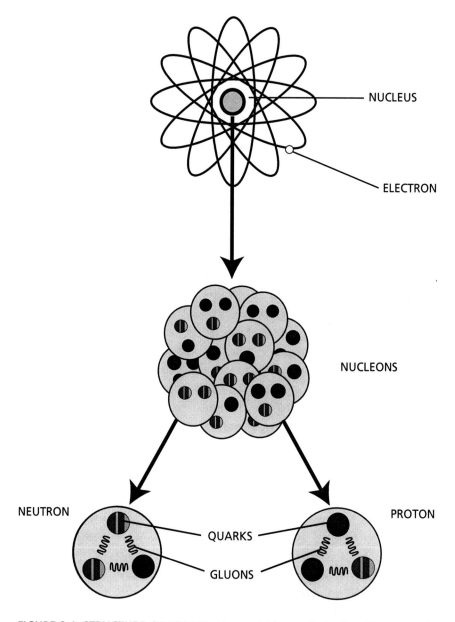

FIGURE 3-1. STRUCTURE OF ATOMS. Atoms which constitute all matter, are made of nuclie and electrons. Nuclei are made of protons and neutrons, collectively called nuclens. Neucleons are made of quarks. A positively charged proton is made of two up quarks (each with +2/3 electric charge) and down-quark (-1/3 charge); the neutron, a neutral particle, is made of two down-quarks. Gluons tie quarks together.

one second old, quarks were formed, and when it became one hundred seconds old, the protons and neutrons bonded to form hydrogen and helium. One billion years after the Big Bang, the galaxies formed.[17] (Figure 3-3)

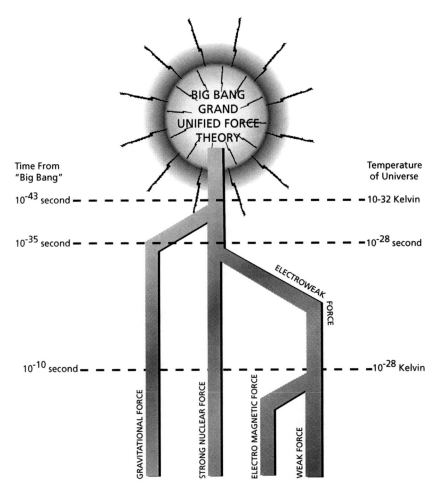

FIGURE 3-2. SYMMETRY BREAKING. A theory or atomic process is said to possess symmetry if it undergoes no change when certain operation are performed on it. Such a state existed when Big Bang occurred. The four forces--gravitational, strong, weak, and electromagnetic--united into one force (Grand Force or Theory). As the ambient temperature cooled following the Big Bang, the Grand Unified Force disassociated into four forces.

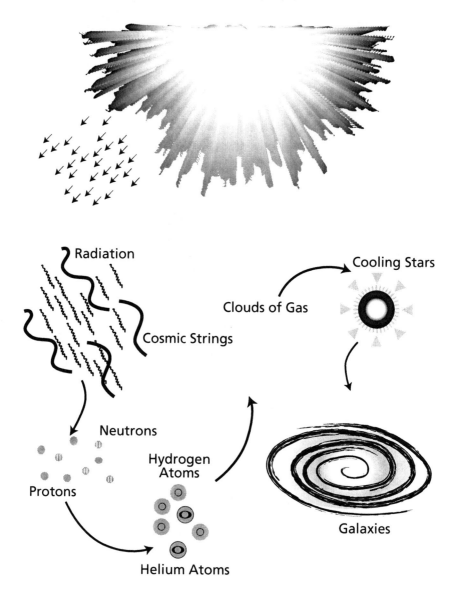

Figure 3-3. BIG BANG

Physics does not have any explanation for the beginning of creation, in which everything in the universe was condensed to a singularity in which the density of matter became infinite and space-time was reduced to a mathematical point. In the singularity, the extreme heat unified all four forces. Similarly, the origin of the four fundamental forces and quarks remains a mystery. Are the four fundamental forces really fundamental? Is the quark the final elemental particle that is indivisible?

After the colossal explosion called the Big Bang, the universe began an expansion that has never stopped. Edward Hubble, who was inspired by the red shift of the galaxies, discovered evidence for an expanding universe in 1932. In 1842, Christian Doppler pointed out that the frequency of the sound we hear is higher when a train is approaching and lower when it is moving away. This is known as the Doppler Effect. The French physicist Armand Fizeau showed that the Doppler Effect works for light as well. The light frequency "red" is longer when the source of light moves away from the observer and shorter when it approaches.

When light passes through a spectroscope, light waves spread out. The longest wave (red) will be at one end, and the shortest wave (violet) at the other end. Light waves between the red and violet waves are based on various frequencies. Therefore, we get the following order of color—red, orange, yellow, green, blue, and violet. Where some wavelengths are missing, we see dark lines on the spectrum. When a star moves away from Earth, the dark lines shift towards the red end of the spectrum. This is called *red shift*. If the star gets closer to the observer, the dark lines shift to the violet end of the spectrum (Figure 3-4). The observation of this phenomenon led Hubble to the conclusion that different galaxies are moving away from each other and that the universe is expanding.[18]

We have reviewed the antiquated history of cosmology that existed prior to the cosmology based on modern physics. We also reviewed the Biblical cosmologies. The Quran, the Muslim's word of God, does not have a chapter on Genesis, but it has verses that deal with Creation. The Quranic concept of the universe differs from that claimed by Herman Bondi, Thomas Gold, and Fred Hoyle, who claim that ours is a steady-state universe that never had a beginning

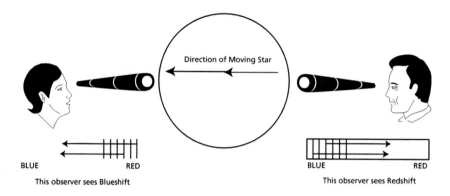

DOPPLER EFFECT

FIGURE 3-4. DOPPLER EFFECT. Radiation from an approaching star is compressed to shorter wavelengths (〰〰) than usual and the dark lines shift to blue color (Blue Shift). Radiation from a receding star is stretched to longer wavelengths (〰〰) than usual and the dark lines shift to red color (Red shift).

and will continue to exist in a condition similar to its present. The universe, according to the Quran, has an origin and an end, as we shall presently discuss.

Knowledge of the external world is reflected in the sophistication of the language that people use. We invent new words to convey new knowledge. When the Quran came to Arabia, the people of that time and place used "the heavens and the earth" to describe the universe. The Quran describes the heavens and the earth (the universe) as a single entity that God split into the heavens (galaxies) and the earth.

> **21:30** Do not these unbelievers see that the heavens and the earth were an integrated mass, then We split them and made every living thing from the water? Will they not believe even then?

The Quran further implies that the universe evolved from a gaseous state that is described as "smoke."

> **41:11** And He (it is who) applied His design to the skies, which were [yet but] smoke; and He [it is who] said to them and to the

earth, "Come [into being], both of you, willingly or unwillingly!"—to which both responded, "We do come in obedience."

Science has come to agree with what the Quran has described in the sixth century. Most astrophysicists concur that the universe began with the Big Bang. The early universe was filled with radiation and elementary particles, such as quarks. Then, hydrogen and helium were formed from the elementary particles in the inferno of the Big Bang. They believe that the gas clouds condensed and gave birth to stars. These eventually took shape to form the galaxies.[19]

In spite of this convergence of the scientific perception of the origin of the universe and the literal Quranic concept of Creation, many Muslims of our time are afraid that the Big Bang theory could be replaced by another paradigm; and that if that happened, their faith might be weakened. Some modern Islamic scholars such as Muhammad Asad argue, "It is, as a rule, futile to make an explanation of the Quran dependent on scientific findings, which may appear true today but may equally be disproved tomorrow by new findings."[20] Whether or not the Quranic verses agree with modern science, these verses describe an explosive beginning from an integrated mass or singularity. The Quran also describes the universe as one that is expanding:

> **51:47** And it is We who have built the universe with [our creative] power; and it is We who are steadily expanding it.

The Quran reveals many signs and events that would take place before the universe ends. The following verse states that the end is heralded when the *currently* "receding stars" and planets are commanded to reverse the direction of their courses:

> **81:15–16** So, I call the *receding stars* to witness, the planets withdrawing into themselves.

The verb (*khunasa*), used in the Arabic text of this verse, means "to recede."[21] The Quranic statement of the reversal of the direction of the movement of the stars is evidence that the universe is currently expanding. Here, too, astrophysicists seem to agree with the Quran when they state that following the huge explosive beginning, the universe began an expansion that is still ongoing today.[22]

The majority of astronomers generally agree on how the universe began, and the evidence from the phenomenon of red shift as described above serves as the basis for the scientific assumption that the universe is expanding. Their views differ, however, on the fate of the universe. Astronomers have two theories on the subject. One holds that the expansion of the universe is not self-limited and will continue. The galaxies will fly away from each other, space will become emptier, and eventually some stars will run out of nuclear fuel. They will shrink, cool, and end up as black dwarfs. Some stars will die as supernovae, exploding and releasing tremendous energy. The debris of the explosions will crumble inward due to immense gravity, to form neutron stars. The heavier neutron stars will again be crushed and shrunk due to gravity until they finally become invisible black holes. Nobody knows how black holes die.

The other view rejects the above scenario and holds instead that the expansion of the universe will come to a halt at some point in the future. When this time comes, the outward momentum of the primordial explosion and the inward pull of gravity will momentarily hold the universe in a balance. Then, gravity will pull the components of the universe together at an increasingly faster pace until the universe collapses into the heat and chaos from which it emerged a few billion years ago.[23] Some astronomers believe that the universe will not come out of the collapsed state, while others speculate that it will rebound with a new explosion, experience a new moment of creation, and "become an entirely new world, in which no trace of the existing Universe remains."[24]

The Quran acknowledges an end to the universe; the fundamentalist Christian view of the fate of the universe does not. The Bible presents the destiny of creation as eternal in Psalms 148:3-6 state:

3. Praise ye him, sun and moon: praise him, all ye stars of light.
4. Praise him, ye heavens of heavens, ye waters that be above
5. Let them praise the name of the LORD: for he commanded, and they were created.

6. He hath also stabilized them *forever and ever: he hath made decree which shall not pass.*

Gerald L Schroeder, an applied physicist and a theologian who received a doctoral degree from Massachusetts Institute of Technology, comments:

> "Based on these verses, sages have said that the universe can exist forever. But if its destruction is to come, it will not be a natural act. It will be a special force from God as was its creation. The heavens are eternal. This is true according to the Bible's description of our universe and according to the best estimate that science has for the situation."[25]

The following excerpt from "Did Man Get Here by Evolution or by Creation?" a publication of the Watch Tower Bible & Tract Society, suggests that our universe will last forever:

> "With bad conditions and violently wicked men no longer existing, the whole earth will become the possession of people who love righteousness. The Bible states: 'The meek ones themselves will possess the earth, and They will indeed find their exquisite delight in the abundance of peace. *The righteous themselves will possess the earth, and they will reside forever upon it.'*—Psalm 37: 11, 29. That is how God will more than compensate for the wickedness that innocent people have suffered during their lifetime. Throughout eternity, God will shower down blessings on mankind, so that any hurt they have received in the past will fade to dim memory, if they care to remember it at all. The Creator guarantees: 'I am creating new heavens (a new heavenly government) and a new earth (righteous human society); and the former things will not be called to mind, neither will they come up into the heart. But exult, you people, and be joyful forever in what I am creating.'—Isaiah 65: 17, 18.30[26]

Thus, the literal interpretation of Jehovah's Witnesses regarding the fate of the universe is that after the second coming of Jesus, the wicked will be destroyed and the righteous will inherit and "reside forever upon" the earth. If the righteous are going to stay forever on the earth, the earth has to be eternal, too.

While the scientific data on the fate of the universe is derived from piecemeal experiments and observations by modern scientists, the Quranic description of the fate of the universe can be deduced from its account of the events that will take place on the Last Day (*Yawm al-Qiyamah*). It has been explained earlier (see chapter 1) that the Last Day should not be conceived of as the earthly twenty-four-hour day but as an aeon or a long period of time. The descriptions of the events on the Last Day are scattered over many chapters and verses in the Quran. The following verse indicates that the universe, including the earth, has an "appointed term" for existence:

> **46:3** We created not the heavens and the earth and all between them but for just ends, and for a term appointed.

The Quran describes the events of the Last Day, but it does not explain how these events come about. We know, however, that they follow the physical and chemical laws of the universe. We have seen earlier that the Quran asks the believers to *"journey in the land, then behold how God originated creation."* (29:20) According to this verse, the products and method of creation are understandable. Therefore, the process and events at the end of the universe are also understandable.

It is tempting to quote verses from the Quran on the events of the Last Day (*Yawm al-Qiyamah*) because quotes from modern scientists appear to repeat verbatim pertinent Quranic verses. One scientist claims that "in some tens of times 10^9 years, the recession of the neighboring galactic systems will cease, and the process will start to reverse."[27] Another scientist contends that the universe will start contracting and that "at first the contraction will be slow, but over many billions of years, the pace will accelerate."[28] The following Quranic verse seems to introduce, in the common layman's language of the Arabs during the sixth century, the scientific facts of our age:

21:104 On that day when We shall roll up heaven as a scroll for writings; As We originated the first creation, so We shall bring it back again—A promise binding on Us; So We shall do it.

The above verse describes the universe as a written scroll of God. The metaphor of the scroll seems to be a figurative explanation of the collapsing universe of science. If we release an open scroll, it will recoil slowly in the beginning and as it rolls back towards the end, the process accelerates. After the Big Bang, the galaxies began to recede. "The galaxies that are now receding from one another will start to approach each other, gathering speed all the time."[29] Over fifteen centuries ago the Quran stated that the receding galaxies would reverse their courses on the Last Day. Fourteen centuries ago, the Quran stated in the chapter, *Coiling Up* [of the universe], that the receding galaxies would reverse their courses before the end of the universe:

81:15-16 So verily, I call to witness the stars that recede.

Likewise, as the result of the final "rolling up" of the universe, astronomers would see the stars advancing (blue shift) toward them rather than receding (red shift), as they see today. During the final collapse, what happens to our solar system? Astronomers believe that under the combined influence of hydrogen fusion on the solar surface and the high temperature helium fusion in its interior, the exterior of the Sun will inflate and cool.[30] This cooling of the Sun may be what the Quran alludes to in a verse that describes the Last Day:

81: 1 When the Sun shall be darkened.

According to scientific interpretation, after the expansion and exterior cooling, the Sun will become a red giant star. When the Sun enlarges, it will swallow the planets Mercury and Venus, and probably the earth as well. The inner solar system will then becomes a part of the Sun.[31] The following verse seems to describe the red bloated sun:

70: 8 Upon the day when heaven shall be as molten copper.

The Quran speaks of a merger between the moon and the sun during the last days of the universe:

75:9 And the sun and moon are united

It does not mention other planets of the solar system, because the common Arabic language at the time of revelation did not include the words Mercury and Venus. However, all Arabs definitely knew the moon. Rotation of the moon was the basis for their calendar. According to science, when the collapse of the universe proceeds uninterrupted billions of years from today, the sun will gradually become red and inflated. Its overwhelming heat will melt the Artic and Antarctic ice resulting in coastal flooding throughout the world.[32] Similarly, the Quran states that the last days of the universe will be heralded by rising water levels in the oceans and by flooding along the coasts:

52:6 And the ocean filled with swell

The Arabic verb *sajara* in the text of this verse means, "to fill."[33] According to current cosmology, flooding of the continental coasts is the result of the "filling" of the ocean due to the melting ice of the poles. As the collapse continues, the soaring ocean temperature will lift clouds of steam into the air. Finally, the unforgiving extreme heat from the sun will make the oceans boil.[34]

The Quran seems to be ahead of its time in warning us that during the last days of the universe, the ocean will boil:

81:6 …when the sea shall be set boiling

Science tells us that during the collapse of the universe, the "observer would no longer be able to discern individual galaxies, for these would now have begun to merge with each other as intergalactic space closes up."[35] This quote is similar to the Quranic revelation that the observable shining points in the sky will decrease during the merger of the galaxies such that the sky appears empty:

81:11 …when the sky stripped bare[36]

Science tells us that as the end of the current universe approaches, the galaxies and subsequently the stars will collapse and merge into each other until a cataclysmic inferno ensues. The continuing collapse of the universe will result in a singularity identical to where our universe originally began.[37] The Quran, in the following verses, seems to narrate the events of the Last Day like a modern astronomer describing the events of the vanishing universe:

25:25 The day the heaven shall be rent asunder with the clouds… (The "clouds" may be those originating from the boiling ocean.)

82:1–2 When the sky cleft asunder, when the stars are scattered

81:2 When the stars fall, losing their luster

69:14 And (on that day) the earth is moved, and its mountains, and they are crushed to powder at one stroke.

55:37 When the sky is rent asunder and it becomes red like ointment

36:53 It will be no more than a single blast…

The events that the Quran describes seem to fit the formation of black dwarfs, supernovae, and the final event of the universe collapsing into a singularity. What happens after the big crunch? When the collapse of the universe is complete, all the matter in the universe will be compressed into a single point. At this point of infinite space-time curvature, every atom and every particle in our universe is crushed out of existence.[38] This point becomes a state of singularity—a single point of space and time, where density and temperature become infinite and theories of science become invalid. One might speculate that there were new laws that worked at singularities, but, according to Stephen Hawking, it would be very problematic even to formulate such laws at such unruly points, and our observations would not help us to define what those laws might be.[39]

Whether the universe rises again phoenix-like from the singularity is not known.[40] Should the universe rise again, Carl Sagan asks: what would be the nature and properties of the new universe? Would the laws of physics and chemistry governing the new universe be different? In the universe that springs out of the Big Crunch, would there be stars and galaxies and world, or something quite different?[41] One wonders whether the previously cited ayat describing the scroll answers the above questions:

> **21:104** The day when We shall roll up the heaven as a scroll is rolled for writings; as We originated the first creation, so *We shall bring it back again*—a promise binding on Us; So We shall do.

The literal meaning of the verse above suggests that a new universe will spring out of the singularity. Similarly, the following verse could be the answer to Carl Sagan's questions.

> **14:48** On the day when the earth is changed into a different earth and heavens into new heavens, Mankind shall stand before God, the One, who conquers all.

The verse suggests that, after the Big Crunch, the newly created universe will be based on new laws of physics and chemistry and maybe different mathematics. Two plus two may not become four. Also, the new universe of the Hereafter would not have time-space relationship as we have seen earlier.

In summary, we find much agreement between Quranic revelation and scientific discovery. Both maintain that the universe originated from nothing and that it evolved into its present state over a long period of an earthbound time. They agree that the universe is expanding and that at some remote future, it will collapse into a single entity or singularity where existing physical and chemical laws become nonexistent, mathematics do not add up, and numbers become infinite. Not only the Quran but also some scientists believe in the rise of a new universe. The question of whether or not a new universe will be created after our current one collapses remains controversial. Scientists who grew up with Judeo-Christian traditions cannot rationally concur with the Biblical description of the origin of the universe, so they vehemently oppose the idea of the supernatural origin of the universe. For Muslims, however, the modern discoveries of science are less problematic.

CHAPTER 4

The Age of the Universe: Uniformitarianism, Fossils, and Muslims

A PREVALENT PUBLICLY held assumption involves the idea that Jews and Muslims share the Christian fundamentalist view in the debate between science and creationism. Western scientists' lack of knowledge about Islam prompts them to preach scientific "truths" to all believers, including Muslims. For example, Maitland A. Edey, the editor-in-chief of Time-Life Books, and Donald C. Johanson, director of the Institute of Human Origins in Berkeley, California, proselytize:

"We talk today about 'Gospel truth' as the truest kind of truth, the kind we swear by, the kind we know in our bones to be true, the last truth of all that we are willing to give up. It was that kind of truth, a universal belief shared by all traditions, all prejudices, all faiths, that had to be unraveled and restated. . . . Indeed, if we truly believe in God, we should recognize that He gave us the brains to conduct scientific research. Not to do so—not to use the

marvelous gift of intelligence with which He has endowed us—
would seem to be disrespectful of God."[1]

If Muslims of our time followed what the Prophet Muhammad
and the Quran command them to do, the above authors could have
limited their admonishments to the Biblical literalists among Jews
and Christians. The Prophet said:

> "Seek knowledge; for its acquisition is fear of God, its pursuit is
> worship, its discussion is prayer, search for it is holy war, its propa-
> gation is charity, its teaching is fraternity. For it is the index of
> right and wrong; it is the lighthouse of the road to paradise. It
> provides consolation in loneliness, friendship in estrangement, fi-
> delity through thick and thin, a protective arm against enemies,
> rapprochement towards foreigners, decor among friends [and]
> through knowledge, God is obeyed and worshipped. In knowl-
> edge the good is revealed, loved and implemented, for it is the
> prius of action, the first principle of every good deed."[2]

Few Western historians have acknowledged the above belief,
which was an integral part of the faith of early Muslims. J. Bronowski
states, "Muhammad firmly emphasized that Islam was not to be a religion of miracles. It became in intellectual content a pattern of contemplation and analysis."[3] This Islamic doctrine paved the way for
the development of modern science and the Western Renaissance.
The common Western belief that scientific thought originated in the
West constitutes a misrepresentation of historical fact. Dr. Robert
Briffault, a British historian acknowledges:

Roger Bacon learned Arabic and Arabic science. Neither Roger
Bacon nor his namesake has any title to be credited with having
introduced the experimental method. Roger Bacon was no more
than one of the apostles of Muslim science and method to Christian
Europe, and he never wearied of declaring that knowledge of Arabic
and Arabian science was for his contemporaries the only way to true
knowledge.[4]

Accounts written by Western historians of science credit Lyell and
Hutton for the theory of uniformitarianism, Leonardo da Vinci for

the discovery of the origin of fossils, and Charles Darwin with recognizing that the origins of humanity from the apes and the phenomenon of natural selection. To date Western historians have not given credit to brilliant Muslim scholars such as al-Biruni, Ibn-Arabi, Ibn-Khaldun, al-Jahiz, Ibn-Sina, Ikhwan al-Safa, Ibn Haithem. On the contrary, they claim that Judeo-Christian and Muslim scriptures teach that the Earth and universe as a whole are very young, and proclaim that Western scientists originated the theories enumerated above. Let us examine the validity of their claims.

Table 4-1.
THE BIBLICAL GENEALOGY OF ADAM TO ABRAHAM

Name	Date of Birth after Creation of Adam	Length of Life in Years	Date of Death after Creation of Adam	Source in Genesis
Adam	930	930		5:3, 4
Seth	130	912	1042	5:8
Enos	235	905	1140	5:11
Cain	325	910	1235	5:14
Mahalaleel	395	895	1290	5:17
Jared	460	962	1422	5:20
Enoch	622	365	987	5:23
Methuselah	687	969	1656	5:27
Lamech	874	777	1651	5:31
Noah	1056	950	2006	9:29
Shem	1556	600	2156	11:10,11
Arphaxad	1658	438	2096	11:14,15
Eber	1723	464	2187	11:16,17
Peleg	1757	239	1996	11:18,19
Reu	1787	239	2026	11:20,21
Serug	1819	230	2049	11:22,23
Nahor	1849	148	1997	11:24,25
Terah	1878	205	2083	11:32
Abraham	1948	175	223	25:7

Western historians maintain two views on the age of the universe: the Judeo-Christian and the scientific. Until the mid-nineteenth century, the West knew only what the Bible revealed. The Book of Genesis furnishes genealogical data in chapters 4, 11, 21, and 25. In addition, the Hebrew Bible gives durations for such variables as how long a particular king reigned, the age of a person when a son was born, and how many years elapsed from one event to another. Based upon this Biblical genealogy, Abraham was born 1948 years after the creation of Adam (Table 4-1).[5 & 6]

In addition to the above genealogy, some Biblical passages describe events to which other historians assign firm dates. This means that we can carefully work our way backward and perhaps calculate the year of the events with which the Old Testament begins. One person who did this rather early on was an Irish-born Anglican bishop, James Ussher (1581-1656), who concluded that Adam and Eve were created by God in 4,004 B.C.[7] Others have improved on this, "pinpointing the actual moment of human creation at 9 a.m. on Sunday, October 23."[8] By adding the years between the creation of Adam and the birth of Jesus (4,004 years) to the years between the birth of Jesus and our time (2000), strict creationists estimate that the universe was created roughly six thousand years ago. Some among them allow an age of ten thousand or even twenty thousand years for the universe.[9]

The pre-Renaissance Jewish scholarly calculation of the age of the earth is no different. The earliest portions of the Hebrew Bible were compiled when Jews were captives in Babylonia (586-539 b.c.). They incorporated the Babylonian version of primeval history, including the story of the worldwide Flood. By tracing the long lifespan of the antediluvian patriarchs and their sons, one might calculate the year of the creation of Adam and Eve. According to Jewish scholars, God created the universe in 3760 b.c.[10]

Just over a millennium ago, in 1000 a.d., Abu Raihan al-Biruni, a Muslim scientist, recorded the methods used by Jews and Christians to calculate the year of creation. One method was as described above. Al-Biruni described another method whereby Jewish and Christian scholars used the total numerical value of particular words in the Hebrew Torah or Syriac version of the Bible respectively, to calculate the years between Adam and Alexander the Great. He says: "The Jews and Christians differ widely on this subject; according to the doctrine of Jews, the time between Adam and Alexander is 3,448 years, whilst according to the Christian doctrine it is 5,180 years."[11] We can calculate, therefore, by adding the years from Adam to Alexander plus the years from Alexander to 2,000 a.d., and reveal that until recently the year for creation of the universe was 5,771 years ago for many Jews and 7,506 years ago for many Christians. Many post-Renaissance Christians and Jews reject the calculation of the age of the earth based on Biblical genealogy. They claim that such calculations are not reliable because the genealogy is incomplete and documents only important people. They do not cite any references from antiquity to support their argument, however. If they have a reference, it is invariably from Jewish scholars who lived among Muslims.

Large numbers of Americans in our technological era still believe that the earth was created very recently. A late 1970s Gallup Poll found that 42 percent of the public believed that the Bible is literally true. According to the same poll, in 2007, the number of literal believers was about one third of American adult population.[12] Another national poll in 1982 reported in the *New York Times* found that 44 percent of Americans agreed that God created man pretty much in his present form within the last 10,000 years. These polls indicate that creationist beliefs are not an extremist view, as nearly half of the American population believes in them.[13] When scientific knowledge about the ancient nature of the earth, the theory of evolution, and uniformitarian principles (discussed below) all started to spread in the Judeo-Christian West, new theological doctrines emerged to accommodate science.

(These include the old-earth creationism based on the gap, day-age, poetic, and revelatory theories that we reviewed in the previous chapter.)

Despite the rise of new interpretations, almost half of modern-day Americans continues to insist upon the recent origin of the universe, even though many contemporary Christian leaders argue for an accommodation of their faith with the modern science after taking into consideration the scientific evidence. Many Christian leaders, including Pope John Paul II, have stated that "the kind of strict

literalism implicit in Ussher's approach is in no way a necessary con-
comitant of religious faith."[14]

One cannot subscribe to the idea of a six- or seven-thousand-year-
old Earth for numerous incontrovertible reasons. For example, ob-
servance of a cross section of a tree shows an annual variation in its
growth indicated by a series of concentric rings, one for each year of
growth. The number of rings in the cross section is equal to the age
of the tree. (Figure 4-1) The oldest known trees are the bristlecone
pines. A study of these pines and their concentric rings shows that
some of these trees are approximately eight thousand years old. If we
accept Bishop Usshur's calculations, they are much older than the
Earth.[15]

The speed of light is 186,000 miles per second. The distance that
light travels in a year is a *light year.* It takes eight minutes for light to
reach Earth after it leaves the Sun. The *Voyager's* signals from Saturn
reached Earth more than an hour and half after they were trans-
mitted. The stars we see today are tens of thousands of light years
away. In fact, we see them as they were millions of years ago, when
their light left them to travel toward us. Therefore, we must concede
that these stars are several million years older than the six-thousand-
year-old universe. The late George O. Abell, professor of astronomy
at the University of California, asks, "Unless, that is, we are to suppose that God created all those remote stars less than six thousand
years ago and at the same time created light from them already well
on its way to us. Why should God do this?"[16]

Another contradiction to the belief in a six-thousand-year-old universe is the continental drift, which is the large-scale horizontal displacement of continents relative to each other.[17] At one time all the
continents were a single landmass, but they gradually drifted from
Africa (Figure 4-2). As scientist George O. Abell explains: "We can
now directly measure the motion of the continents. Laser satellite
experiments show that North America and Europe are separating
at about two centimeters per year [and] to separate the twenty-five-
hundred-mile beach across the Atlantic at that rate (less than one
inch per year) has taken approximately 200 million years."[18]

We have established that the universe is older than six thousand
years. We have also seen the variable opinions among the Jews

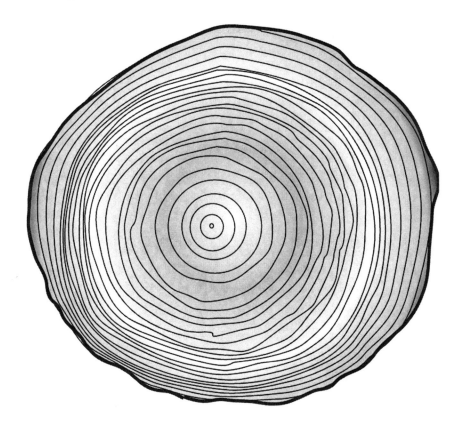

FIGURE 4-1. CROSS SECTION OF A TREE TRUNK. Note the annual rings of alternating light and dark wood. The xylem fomed during each growing season forms distinct bands. They are called the annual rings, which can be counted to determine the age of the tree.

and Christians regarding the age of the universe. Contemporary Muslims, on the other hand, do not entertain possible interpretations of this issue. That is not surprising because since the beginning of the Muslim Dark Ages they were not allowed to think for themselves. Only Muslim scholars (*ulamah*) were permitted to do so. Most Muslims believe that the *ulamah* are the only ones capable of interpreting the Quran and providing the world with the truth. Any opinion regarding spiritual or secular matters that does not correspond to that of the *ulamah*, or other individualistic attempts at interpreting the Quran, is considered a heresy. Another reason Muslims

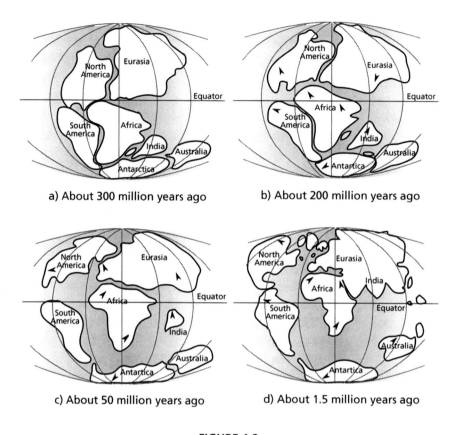

a) About 300 million years ago

b) About 200 million years ago

c) About 50 million years ago

d) About 1.5 million years ago

FIGURE 4-2

lack knowledge of the age of the universe is the rampant ignorance and illiteracy among them.

The third reason is lack of freedom of expression in the contemporary Muslim world. Muslim rulers have bottled up the Muslim mind. Alas! The West now imagines and redefines that bottled-up mind in their books, newspapers, and magazines while Muslims at large remain in the bottle as mere spectators with no voice of their own. Those of us in America who have the freedom to reflect about the universe should ask: What is the age of the universe based on the literal reading of the Quran and based on Muslims who lived prior to the beginning of fourteenth century?

The Quran states that God created the universe in six days (*fi sitat-i ayyam*): "Your Guardian—Lord is Lord, who created the heavens and earth in six days..."(7:54) The verse refers to a process that began before the earth and sun were created. It places the creation of the universe on a time scale. As discussed in chapter 2, the six days referred to above are not equal to six earthly days: "Verily a day in the sight of thy Lord is *like* a thousand years of your reckoning. (22:47) The thousand years mentioned in the verse are not *equal* to a thousand years of our earthly time. The six days of creation means six long periods of earthly time. Even though we cannot precisely determine the age of the universe from the above Quranic verses, we can reasonably conclude that the universe was created a very long time ago. Does this inference agree with the opinion of early Muslim scientists?

Seven hundred years before James Hutton (1722-1799), the earliest Western author of uniformitarianism, Abu Raihan al-Biruni (973-1048) was born in Khiva, Uzbekistan. Al-Biruni was among the most educated and cultured men of his time, with a brilliant intellect "encyclopaedic in scope."[19] In addition to mastering the Arabic language in which he wrote, al-Biruni was proficient in Turkish, Persian, Sanskrit, Hebrew, and Syriac. He excelled in numerous fields of knowledge and particularly in astronomy, mathematics, chronology, physics, medicine, and history. He died in Ghazani, Afghanistan.

Al-Biruni believed that the age of the earth and universe was so great that he could not speculate on it. He criticized the then held Judeo-Christian teachings of the six days of creation in Genesis as equal to the six earthly days of twenty-four hours each. In *Kitab Tahdid Nhayat al-Amakin*, he states:

> "I say: if by adducing rational proofs and true logical syllogisms, we can conclude that the world was created, and that the parts of its finite period, since its creation and existence, have had a beginning, we cannot by such proofs deduce the magnitudes of those parts, which will enable us to determine the date of the creation of the world.[20]

Al-Biruni points out that while Jews, Christians, Sabians, Magians, and others agree about marking events in relation to the "era" during which mankind was created, they do not agree in estimating the duration of that era. Al-Biruni clarifies that while they consider the day of Genesis 1:1, 2 as "the first day of the week in which the world was created," he rejects this Judeo- Christian interpretation of the Genesis because he believes that days of creation "cannot be measured by a day and night." The reason for his rejection of this belief is that the cause of day and night is the sun "rising and setting," and that "the sun and moon were created on the fourth day of the week." He asks, "How is it possible to imagine that these days are like the days of our reckoning!" Then he counters these Judeo-Christian claims by quoting from the Quran: "A day in the sight of thy Lord is like a thousand years of your reckoning" (22: 47) and "In a day the measure whereof is as fifty thousand years" (70: 4). Al-Biruni clearly believes that "we cannot estimate that period [of the creation] with our method of reckoning, and it is unverifiable since the beginning of creation." He used the Quran and logic to support his argument.

Moreover, al-Biruni stated a thousand years ago that God initially scattered the celestial bodies and that the time between creation and the present could be billions of years. He states:

> "For it is quite possible that these (celestial) bodies were scattered...when the Creator designed and created them. If you then ask the mathematician as to the length of time, after which they would meet each other in a certain point, or before which they had met each other in that identical point, no blame attaches to him, if he speaks of billions of years."[21]

The "scattering of the celestial bodies" can be explained better with our modern concept of "the Big Bang and expansion of the universe."

Ameer Ali, a contemporary historian, described the close of the tenth century as the darkest period in the history of Islam in its rejection of rationalism and science. In his words, a "heartless, illiberal, persecuting formalism dominated the spirit of theologians; a pharisaical Epicureanism had taken possession of the rich and an ignorant fanaticism of the poor; the gloom of the night was fast thickening, and Islam was drifting into the condition into which ecclesiasticism had led Christianity."[22] During this epoch of travail and sorrow for all lovers of truth, a small body of thinkers joined each other to form a brotherhood aimed at keeping aflame the torch of knowledge among Muslims. They called themselves Ikhwan al-Safa or the Brethren of Purity. We know little about their membership with the exception of the following: Abu Sulayman al-Busti (known as al-Muqaddasi), Abu'l-Hasan al-Zanjani, Abu Ahamed al-Nahrajuri (known as al-Mihrajani), al-'Aufi, and Zaid ibn-Rifaa.[23] The tenth-century *Epistles of Ikhwan al-Safa* (*Rasa'il Ikhwan al-Safa*) is not the work of any one member but a collaborative effort.

The *Rasa'il* is comprised of fifty-two epistles in which knowledge is classified into three classes: mathematics, physics, and metaphysics. The brethren maintained that the earth was created a very long time ago, and therefore has a long history. They tell us that the individual cycles of the stars affect events on earth and that it takes thirty-six thousand years for all the heavenly bodies to complete one cycle. Some astronomical revolutions and conjunctions *Rasa'il Ikhwan al-safa* describes occur only once in a long time, while others repeat more frequently. "A very long period is that of revolution which occurs only in 36,000 years." Every three thousand years the fixed stars, apogees, and nodes of the planets pass from one astrological sign to another after having traversed all the degrees of these signs. Every nine thousand years they pass from one quadrant to another, and as the effect of their rays at diverse points of the earth are modified, so are the climates of different countries. Every thirty-six thousand years the stars conjunct and the planets reunite at the degree of the sign of Ram to begin a new cycle.[24] If it takes thirty-six thousand years to make one cycle, clearly Muslims accept that universe (and the Earth) is much older than the Biblical six thousand years.

Abu Ali ibn Sina (Latin name: Avicenna) (980-1037 A.D.) is one of the most "famous and influential of the philosopher-scientists of Islam."[25] His theory about the origin of mountains confirms the ancient age of the earth. He gives two causes for the emergence of mountains:

"Either they are the effects of upheavals of the crust of the earth, such as might occur during a violent earthquake, or they are the effect of water, which, cutting for itself a new route, has denuded the valleys, the strata being of different kinds, some soft, some hard. The winds and waters disintegrate the one, but leave the other intact. Most of the eminences of the earth have had this latter origin. It would require a long period of time for all such changes to be accomplished, during which the mountains themselves might be somewhat diminished in size.[26]

Ibn Sina's statement, "It would take a long time for all such changes," is another proof that Muslims never entertained the Judeo-Christian belief of a young earth.

If we are to believe that the earth is only six thousand years old, as inferred in Biblical genealogies, then we should declare theory invalid.[27] Charles Darwin could not reconcile the theory of evolution with the Biblical view of a six-thousand-year-old Earth, so he gradually drifted away from the Christian belief and become a materialist. He found compatibility between the theory of evolution and the idea of an antique Earth as stated in Charles Lyell's *Principles of Geology*. Hence, principles of uniformitarianism presented by Lyell should be included in any discussion of the theory of evolution. The West claims that it was Charles Lyell who first declared that "most of the earth's rocks were formed over a long period of time by natural processes observable to this day"[28] and that he had arrived at this conclusion based on James Hutton's four principles of uniformitarianism:

The uniformity of law states that the laws of nature that operate today operated in the past. The assumption about the unvarying natural law serves as a necessary warrant for extending inductive inference into an unobservable past. The uniformity process necessitates explaining the past through causes that are now in operation. The uniformity rate, or gradualism indicates that the pace of change is usually slow, steady, and gradual and that localized catastrophes such as floods, earthquakes, etc. occur. The law of the uniformity, or non-progressionism states that the earth operates in a dynamically steady state. We can infer its past not only from its laws and rate of change, but also from its current order of events. Land and sea might change their positions, but they exist through time in roughly the same proportion. Species come and go, but the mean complexity of life remains forever constant.[29]

The question is whether Hutton and Lyell were the earliest men who understood these principles of uniformitarianism. In spite of the "ballads" written by Stephen Jay Gould and others in the West praising these men for the so-called discovery of the principles of uniformitarianism, a study of the works of Muslim scientists—Ibn Sina, al-Biruni, Ikhwan al-Safa, Ibn-Khaldun, others—reveals that these Muslim scientists understood and introduced these principles long before Hutton and Lyell. Al-Biruni, the foremost Muslim geologist, was a uniformitarian. He identified the Ganges Plain in India as sedimentary deposit. He wrote:

If you see the soil of India...if you consider the rounded stones found in the earth, however deep you dig, stones that are huge near the mountains and where the rivers have a violent current, stones that are smaller size at a greater distance from the mountains and where the streams flow more slowly, the stones that pulverized in the shape of sand where the streams begin to stagnate near their mouths and the sea . . . you can scarcely help thinking that India was once a sea, which by degrees has been filled by the alluvium of the streams.[30]

Moreover, al-Biruni wrote:

We have to rely upon the records of rocks and vestiges of the past to infer that all these changes should have taken place in very, very long times and under unknown conditions of cold and heat: for even now it takes a long time for water and wind to do their work. And changes have been going on and observed within historical time.[31]

Al-Biruni did not invent extinct or unknown causes as explanations for the formation of the Ganges Plains in India. We infer from his statement "however deep you dig" that he used the scientific method to reach his conclusion. We see that he used forces that operate today to explain changes in stones (the uniformity of law and uniformity process) and that he believed in the uniformity rate because he states, "by degrees has been filled by the alluvium of the streams." Furthermore, he recommended the method of observation of rocks and fossils in order to decipher the long history of the earth.

In spite of this affirmation of the gradual and slow character of natural processes acting on the surface of the earth, al-Biruni knew that natural cataclysms and disasters devastated the earth from time to time. Like most other Muslim scientists, he believed that these catastrophes occurred in localized areas. According to al-Biruni:

"The disasters, which from time to time befall the earth, both from above and below, differ in quality and quantity. Frequently [the earth] has experienced one so incommensurable in quality or in quantity or in both together, that there was no remedy against it so that no flight or caution was of any avail. The catastrophe comes on like a deluge or an earthquake, bringing destruction either by breaking of the surface, or by burning by hot stones and ashes that are thrown out, by thunderstorms, by landslides, and typhoons; further by contagious and other diseases, by pestilence, and more of the like. Thereby a large region is stripped of its inhabitants, but then after a while, after the disaster and its consequences have passed away, the country begins to recover and to show new signs of life, then different people flock together like wild animals, who formerly were dwelling in hiding-holes and the tops of the mountains. Assisting each other against common foes, wild beasts of men, and furthering each other in the hope for a life in safety and joy civilizes them. Thus they increase to great numbers; but then ambition, circling round them with wings of wrath and envy, begins to disturb the serene bliss of their life."[32]

The Universal Flood of the Judeo-Christian faith is not a Muslim doctrine, and with the aid of the scientific method, al-Biruni rejected a worldwide flood of the Judeo-Christian description. Muslims do believe it was a localized flood, however. Eight hundred years before Darwin, al Biruni stated:

"Persians, and great mass of the Magians, deny the Deluge alto-gether ;...In denying the Deluge, the Indians, Chinese, and vari-ous nations of the east, concur with them. Some, however, of the Persians admit the fact of the Deluge, but they describe it in a different way from what is described in the books of Prophets. They say partial Deluge occurred in Syria...but did not extend over the whole of the then civilized world, and only few nations were drowned in it...the traces of the water of the Deluge, and effects of the waves are still present on these two pyramids halfway up, above which the water did not rise."[33]

In the above quotes, al-Biruni uses natural causes only to explain the destruction and recovery of people and civilizations. Unlike mod-ern scientists, earlier Muslims believed that physical and chemical laws instituted at the time of the initial creation by God governed these natural events. Al-Biruni did not believe in the fixity of species as Lyell proposed in *Principles in Geology*, but in the changing com-plexity of life forms. His views on the evolution of life will be pre-sented later with the pre-Darwinian Muslim thoughts on the origin of life and its development.

Ikhwan al-Safa (the Brethren of Purity) were also uniformitarians in their outlook. They believed:

"During the lapse of time of this movement [of stars], civilization in this world of generation and corruption is transported from one quarter of the earth to another. Continents replace seas, and seas come to replace the solid earth. Mountains change into seas, seas into mountains [i.e., the law of the uniformity, or nonprogression-ism]...by the effect of their intermediate cause, the zeniths of the stars and the incidence of their rays at diverse points of the earth are modified and with them the climate of diverse countries..That is why the diverse regions of the earth are modified [i.e., the uni-formity of law, the uniformity process, and the uniformity rate, or gradualism]. The layers of the air are changed above diverse places and countries by which the properties of these layers of air pass from one state to another. It is for this reason that cultivated earth becomes deserts, and deserts become cultivated earth; steppes become the sea and the sea becomes steppes or mountains."[34]

Again we find in the words of Ikwan al-Safa all the components of uniformitarianism. Unlike Lyell, they believed in a transformation of species and kingdoms to more complex life forms. We will learn about their point of view about life later. Ibn Sina's (Avicenna's) view on the origin of mountains shows that this Muslim was also a uniformitarian.

"Mountains have been formed by one (or other) of the causes of the formation of stones, most probably from the agglutinative clay, which slowly dried and petrified during ages of which we have no record. It seems likely that this habitable world was in former days uninhabitable, and indeed, submerged beneath the ocean. Then, becoming exposed little by little, it petrified in the course of ages, the limits of which history has not preserved; or it may have petrified beneath the waters by reason of the intense heat confined under the sea. The more probable of these two possibilities is that petrifaction occurred after the earth had been exposed, and that the condition of the clay, which would then be agglutinative, assisted the petrifaction."[35]

Ibn Sina observes that mountains most probably were formed from agglutinative clay, which slowly petrified "during ages of which we have no record" (i.e., the uniformity rate, or gradualism). We saw earlier that Ibn Sina attributed the flow of water and the earthquakes to the origin of the mountains (i.e., the uniformity process). He assumed these causes acted in the past as they acted during his time (i.e., the uniformity of law).

Ibn Khaldun, statesman, jurist, historian, and scholar, was born in Tunis on May 27, 1332. The love of learning and intellectual pursuit for which his father and grandfather were noted produced in him a rare combination of philosopher and statesman. Ibn Khaldun served as an advisor to various rulers of his time. In 1387 he made the pilgrimage to Mecca, after which he returned to Egypt, where he died on March 17, 1406. His well-known work, *Muqaddimah*

(*Introduction*), serves as the preface to *Kitab al 'Ibar,* which is considered the first attempt to interpret the patterns of changes that govern developments in the social and political order of societies. A rational analytical methodology and an eye for detail marked his approach. Ibn Khaldun diverged from historical narration of events based on prevailing social conventions. He created a philosophy of history that was detached from superstition and based on a critical and scientific examination of events. Thus, Ibn Khaldun pioneered a new science and new terminology in the study of history.[36] Arnold J. Toynbee, an eminent British historian, believed that the *Muqaddimah* was the "greatest work of its kind that has ever been created by any mind in any time."[37]

Ibn Khaldun informs us that Christian genealogists used to believe that black Africans were the children of Noah's son Ham and that their blackness resulted from the curse Noah cast on them. He points out that the story in the Torah declares that Ham's children would be slaves to his brother's descendants, but makes no mention of their color. In his view, attributing the blackness of Africans to Noah's curse discounts the influence that natural forces such as warm and cold climates have on living creatures. He explains:

> "Genealogists who had no knowledge of the true nature of things imagined that Negroes were the children of Ham, the son of Noah, and they were singled out to be black as a result of Noah's curse, which produces Ham's color and slavery of God inflicted upon his descendants.... The black skin common to the inhabitants of the first and second zones [Ibn Khaldun divided northern hemisphere of the earth into seven north zones. The first zone runs along equator and is followed to the north successively by the second through seventh zones.] is the result of the composition of air in which they live, and which comes about under the influence of the greatly increased heat in the south. The sun is at the zenith there twice a year at short intervals. In all seasons, the sun is in culmination for a long time. The light of the sun, therefore, is plentiful. People there go through a very severe summer, and their skins turn black because of excessive heat."[38]

Ibn Khaldun attributes the whiteness of the inhabitants of northern zones to the influence of excessive cold climates that occur as a result of the greater distance between the sun and the zenith in those regions. He also attributes to climactic conditions northern individuals' lack of bodily hair as well as their blue eyes, freckled skin, and blond hair, and warns that Christians should acknowledge the environmental factors and influences that create different features among people. He considered it a mistake for Christians to attribute the traits of whole nations to one of their prominent ancestors. Ibn Khaldun also contends that:

> "[There also is disregard of the fact that physical circumstances and environment] are subjected to changes that affect later generations; they do not necessarily remain unchanged. This is how God proceeds with His servants. . . . And verily, you will not be able to change God's ways."[39]

Ibn Khaldun applied the principles of uniformitarianism when he attributed the composition of the human race to causes that were in operation during his time (i.e., uniformity process). He observed that the hot and cold climates and the position of the sun in various zones of the earth remained constant in time and place (i.e., uniformity of law and the uniformity rate, or gradualism). Lyell believed that species came and went while the mean complexity of life remained forever constant. His opinion was similar to the then-held Judeo-Christian view. Edey and Johanson describe it as follows: "Because of a general acceptance of the Biblical account of Creation, almost everybody took it for granted that species were fixed. They have been created by God in their own shape and could not change."[40]

Ibn Khaldun and other Muslims who preceded Lyell did not believe in the story of Noah's curse that human traits are predetermined. On the contrary, they believed that human species did undergo change. Ibn Khaldun maintained that environmental changes "affect later generations." His earlier statement in which he refers to God as the instigator of change affirms the fact that Muslims, in their classical period, believed that the impact of environmental changes on life forms served as the method by which God created new and varied life forms.

All four components of uniformitarianism are obvious in early Muslim thought, but they did not adhere to Lyell's idea of constancy in the complexity of life. Thus, neither the idea that environmental factors cause changes in life forms nor the idea of uniformitarianism represent pristine Western ideas. These principles were widely accepted among early Muslims who routinely used them in scientific pursuit long before Lyell and Hutton.

Western historians of science, who have claimed virtually all scientific discoveries, have also taken credit for the rational explanation of the origins of fossils. Fossils are important strands of evidence supporting evolution. Most people in the Christian West did not know what to make of them. In the Dark Ages, the Church explained that these fossils were Satan's failed attempts to mimic the work of God in creating life. Asimov attributes the first reliable rationalization of the origin of fossils to Leonardo da Vinci (1452-1519), who claimed that they were remnants of obsolete creatures. According to Asimov, Charles Bonnet (1720-1792), in an attempt to avoid an explanation based on the concept of evolution, promoted the belief that fossils were relics from the time of Noah's flood. In 1770 he spread the concept of catastrophism, by which he held that the entire conglomerate of living creatures on the Earth was destroyed by whole series of catastrophes, with a new creation beginning afterwards. Bonnet argued that the Bible dealt only with the Earth after the last catastrophe. As Asimov observes, however, the discovery of more fossils required the recruitment of more and more catastrophes, which led to the failure of the concept of catastrophism. The origin of fossils increasingly began to be seen more as a process of evolution than as a product of catastrophes.[41]

Asimov's statements that nobody knew the origins of fossils and that da Vinci served as the first individual to render a rational explanation about their origin represent another Western myth. Long before da Vinci, early Muslims knew that fossils were the remains of once-living organisms. Al-Biruni gave his opinion about the origin of fossils in his description of the Arabian Peninsula:

> "This steppe of Arabia was at one time sea, then was upturned so that the traces are still visible when wells and ponds are dug; for they begin with layers of dust, sand and pebbles, then there are found in the soil shells, and bones, which cannot possibly be said to have been buried there on purpose. Nay, even the stones are brought up in which are embedded shells, cowries and what is called 'fish-ears', sometimes well preserved, or the hollows are there of their shape while animals have decayed."[42]

Similarly, Ibn Sina's opinion about fossils can be derived from his description of the origin of mountains. He wrote:

> "It would require a long period of time for all such changes to be accomplished, during which the mountains themselves might be somewhat diminished in size. But that water has been the main cause of these effects, is by existence of fossil remains of aquatic and other animals on many mountains."[43]

Important scientific discoveries rendered by Muslim savants regarding the age of the earth, uniformitarianism, and the origin of fossils have almost been completely obliterated from the sight of the world by Western scholarship. Western historians generally start with the Greeks and skip over Islamic history to the Renaissance in Europe. Their amnesia regarding the contribution of Islamic scientists clearly presents a distorted version of history. Why aren't the contributions of Muslims widely known? Professor John William Draper (1812-1883) of New York University and a contemporary of Darwin, points to persistent efforts by Western scholars to suppress such history:

> "I have to deplore the systematic manner in which the literature of Europe has contrived to put out of sight our scientific obligations to the Muhammadans. Surely they cannot be much longer hidden. Injustice founded on religious rancor and national conceit cannot be perpetuated forever."[44]

Draper's last statement clearly attributes the erasure of Muslim contributions from the Western historical record to religious and racial prejudices. In *The Making of Humanity*, British historian Robert Briffault focuses on the deficiencies of the Western historical record by demonstrating that the European Renaissance occurred "under the influence of Arabian and Moorish revival culture [and that] Spain, not Italy, was the cradle of the rebirth of Europe." He explains that while Europe wallowed in ignorance and barbarism, Muslim cities constituted "centers of civilization and intellectual activity." He asserts "[t]hat a brilliant and energetic civilization full of creative energy should have existed side by side and in constant relation with populations sunk in barbarism, without exercising a profound and vital influence upon their development, would be a manifest triumphs of Cross over Crescent" and that "the reclamation of Spain from Moorish yoke" does not, in his view, erase the relationship between the Islamic culture and Europe, despite "the conspiring of every circumstance to suppress, deform, and obliterate the records of that relation."[45]

Briffault echoes Draper's accusations regarding the scholarly suppression of the truth that was "stubbornly ignored and persistently minimized" even by prominent historians such as Gibbon. Not only does Briffault consider the past representations of Islamic culture of the Middle Ages as inaccurate, but also those of the present. Briffault's resentment of scholarly injustice towards Islamic culture led him to the following insight: "[I]t is highly probable that but for the Arabs modern European civilization would never have arisen at all."[46]

Briffault's statements are evidence of scholarly truthfulness, an effort that almost always requires a degree of ethical courage, and without which scholarly studies will lose their credibility as times change. How fitting it is to quote, in this instance, an old African proverb about history and historians:

> "Until the lions have their historian, tales of hunting will always glorify the hunter."

The story of science has often been presented as the victory of scientific discovery over religious mythology. But another story is

CHAPTER 5

A Brief Review
of the Theory of Evolution

I HAVE CHOSEN TO OFFER a brief review of the theory of evolution for the benefit of those who are unfamiliar with the field. This overview is also essential to an understanding of the places in which the Quran agrees with the theory of evolution and the places in which it does not. The Christian fundamentalists present the theory of evolution as anti-God, primarily to the American public but also to Christians worldwide. Recently some Muslims have adopted the hollow Christian fundamentalist polemics to discredit evolution from their pulpits as well. In this chapter we will discuss antievolutionists' ardent outcries against evolution as well as evolutionists' patient response.

* * *

A bush takes life from a seed and grows into a stem with branches, twigs, leaves, and fruit. Similarly, life originated from organic and inorganic chemical matter, grew into an original "stem cell," and then branched into every kind of life form on the earth. Biologists have constructed a model to explain the origin of life and its growth, a process that can be likened to the development of a bush. This model is the theory of evolution.

When those who are not biologists talk about the theory of evolution, however, they are often confused about the words "theory" and "evolution." The Institute for Creation Research (ICR) at San Diego, the foremost American fundamentalist Christian creationist organization, reinforces the innocent confusion that is prevalent among the general public. Many Muslims also use this confusion to reject evolution. Therefore, we need to explain the scientific meaning of these words.

Professor Stephen Jay Gould of Harvard University, a leading advocate for the theory of evolution, explains the word theory as follows: "In American vernacular, 'theory' often means 'imperfect'—part of a hierarchy of confidence running downhill from fact to theory to hypothesis to guess." Anti-evolutionists therefore argue that evolution is a theory and not a fact, and that it is inferior to fact.

Gould clarifies: "[Evolution] is a theory. It is also a fact, and facts and theories are different things, not rungs in a hierarchy of increasing certainty. Facts are the world's data. Theories are structures of ideas that explain and interpret facts." Theories connect all facts together to form an inherently consistent story. Gould points out, "Facts do not go away when scientists debate rival theories to explain them." According to science, "'fact' can only mean 'confirmed to such a degree that it would be perverse to withhold provisional assent.'"[1] In short, the fact of evolution does not go away when varying theories are offered to explain the process.

Charles Darwin did not use the word *evolution*, but *evolve* appeared as the last word in his work, The Origin of Species. *Webster's Third International Dictionary* describes the word *evolution* as "a process of continual change from a lower, simpler, or worse condition to a higher, more complex, or better state." *The American Heritage Dictionary* explains evolution as "a gradual process in which something changes into a different or better form." Darwin himself understood the word as described in the dictionary.

The contemporary scientific definition of the word *evolution* differs greatly from Darwin's understanding and what is given in standard dictionaries. When common people talk about the theory of evolution, however, they tend to use the meaning given by a standard dictionary. Scientists say that these definitions are simply wrong. Standard biology textbooks define evolution thusly: Evolution can be precisely defined as any change in the frequency of alleles within a gene pool from one generation to the next.[2]

Alleles are alternative variants (different versions) of the same gene. For example, humans can have A, B, or O alleles that determine one aspect of their blood type. A gene pool is the set of all genes in a species or population. For geneticists, a population refers to a group of organisms that interbreed or have the potential to do so. The genetic composition of a certain population exists and evolves over time.

Let us use a frequently cited example of the *Biston betularia* (the peppered moth) to explain evolutionary change. These moths are either light or dark. Primarily, a single gene determines the color of moths. The dark moths constituted two percent of the population prior to 1848, then gradually increased until the 1950s. Hence, in the moth population, there was an increase in the presence of the gene that gave the moth dark color between 1849 and the 1950s. The change in the frequency of dark moths represents a change in the gene pool. Scientists call this change evolution.

According to Robert Shapiro, professor of chemistry at the University of New York, "We do not know how life began on earth. The question poses a greater mystery today than it did in the early 1970s."[3] But the general scientific view is that life somehow evolved through a series of chemical reactions. In the 1920s, Alexander Oparin in the Soviet Union and J.B.S. Haldane in England independently introduced ideas that explained the origin of life on Earth; later, other scientists developed a comprehensive theory that has become the cornerstone of scientific explanation for the origin of life. They maintain that the atmosphere of the earth in the past was much different than it is today. After the earth and the solar system were formed out of the condensation of interstellar materials, the atmosphere had an abundance of hydrogen, methane (a simpler reduced form of carbon), and ammonia (a simpler reduced form of nitrogen). New compounds were formed as a result of the exposure of the early atmosphere to lightning, radiation, volcanoes, and meteors. Rainfalls would then siphon these compounds into oceans and ponds, which are sometimes referred to as "pre-biotic soups." The chemicals in the pre-biotic soup combined and transformed into more complex compounds and ultimately into life itself.

Stanley Miller and Harold Urey in the 1950s provided laboratory evidence for the formation of a number of organic compounds,

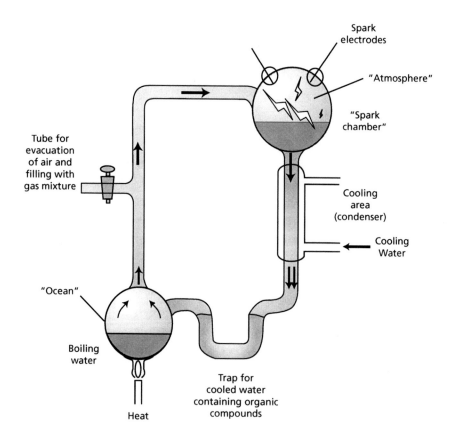

FIGURE 5-1. MILLER UREY EXPERIMENT SIMULATING PRE-BIOTIC CONDI-TIONS. The upper glass chamber ("early atmosphere") has a mixture of hydrogen (H2), metahne (CH4), and ammonia (NH3). Sparks from electrodes represent lighting as a source of energy. The lower glass chamber ("ocean") is boiled to make water vapor, which provides oxygen to the chemical reaction and also carries organic molecules back to the lower chamber ("ocean") by condensation ("rainfall").

including several amino acids, by exposing the mixture of hydrogen, ammonia, and methane to electric sparks for several days.[4] (Figure 5-1). Miller and Robert Orgel assumed that "there must have been a period when the earth's atmosphere was reducing (without free oxygen), because the synthesis of compounds of biological interest takes place only under reducing conditions."[5] The Miller-Urey experiment has been used to support the Oparin-Haldane theory of the origin of life.

If, as Miller and Urey claim, the production of organic compounds depends on the presence of a reducing atmosphere (lack of free oxygen in the environment), then the earth's initial atmosphere must have been reducing to form organic compounds. Recent data points to the absence of such a reducing atmosphere in the past. Professor Shapiro of New York University tells us that, according to geologists and geochemists, the earth's atmosphere was initially neutral or "slightly reducing," and methane and ammonia would not be formed. In addition, radiation from the sun, which recent discoveries have found to be more intense in the past, would have destroyed any reduced gases.

Other information contrary to the Miller-Urey claims includes the fact that no extremely nitrogen-rich carbonaceous deposits have been discovered from which pre-biotic soup could have been formed. Moreover, in 3.8-billion-year-old rocks in Greenland, we see carbon in oxidized carbonate form. Shapiro points to the paradox that the earth, the place where life originated, is a "non-reducing oasis in [a] reducing universe." He concludes that either science has misinterpreted the conditions under which life was generated or that the reducing atmosphere occurred in via a different path.[6]

Philip H. Abelson of the Geophysical Laboratory at Carnegie Institute of Washington maintains that there is no evidence to support the methane-ammonia hypothesis and that if ammonia existed, it would have "quickly disappeared."[7] Erich Dimroth of the University of Quebec and Michael M. Kimberley of Toronto University argue against a reducing atmosphere in the earth's past. They confirm the fact that we find "no evidence in the sedimentary distribution of carbon, sulfur, uranium, or iron that an oxygen-free atmosphere has existed at any time during the span of geological history recorded in well preserved sedimentary rocks."[8]

Henderson Sellers of the University of Liverpool, and A. Benlow and A. J. Meadows of the University of Leicester express a similar opinion. They explain their view as follows:

> "The mean temperature of the early Earth was above 273 K, even though the solar luminosity was low and significant amounts of NH_3 were not present in the atmosphere. Biologists concerned with the origin of life still often quote an early atmosphere consisting of reduced gases, but this seems to stem as much from ignorance of recent advances as from active opposition to them. In the later part of the 1970s, the concept of early-oxidized atmospheres on the terrestrial planets is becoming the new orthodoxy."[9]

After reviewing the concept of the atmospheric state of the primitive earth, Shapiro finds it unscientific that many scientists accept a theory as fact and shun any evidence that contradicts it.[10] Even though there remains much dispute about the origin of life in the earth's early atmosphere, there is no disagreement that nucleotides serve as the building blocks of Deoxyribonucleic acid (DNA) and that DNA is the basis of all life on the earth.

A nucleotide is composed of one five-carbon sugar (deoxyribose), one phosphoric acid radical, and one nitrogen base composed of either adenine, cytosine, guanine, or thymine. Repetition of four units of nucleotides forms a unit of DNA, the basic building block of life (Figure 5-2). Most scientists admit that the "spontaneous" formation of small molecules such as amino acids and nitrogen bases could have occurred under various earthly conditions in the past; however, they are not certain how these molecules could have merged into larger units.[11]

Even though scientists believe that life began in water, until recently there has been no sound explanation for the paradox of polymerization in the presence of water. Polymerization is a chemical reaction that joins small molecules into large chains of molecules. The product formed after polymerization is known as a polymer. Polymerization reactions are generally dehydrations in which a

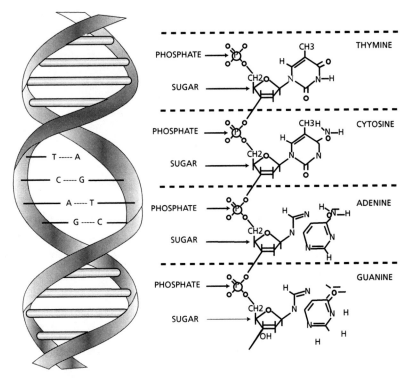

Figure 5-2. Portions of single DNA polynucleotide chain. Each nucleotide is composed of one five-carbon sugar (deoxyribose), one phosphoric acid residue, and one nitrogenous base (adenine, cytosine, guanine, or thymine). The chain also consists of alternating phosphate and sugar residue.

molecule of water is lost during the formation of a polymer. The polymer is unstable in the presence of water.

Shapiro cites A.G. Cairns-Smith, a chemist at the University of Glasgow in Scotland, who maintained that primitive forms of life on earth were made of clay minerals. Cairns-Smith explains that the complex composition of clay minerals facilitates the catalysis of chemical reactions.[12] Carl Sagan summarizes the view of J. D. Bernal, a famous British scientist, as follows:

> "[A]bsorption of molecular intermediates on clay or other minerals may have concentrated these intermediates. Such concentration could offset the tendency for water to break down the polymers

of biological significance. Of special interest is the possibility that such concentration matrices included phosphates, for this would help to explain how phosphates could have been incorporated preferentially into pre-biological organic molecules at a time when biological concentration mechanisms did not yet exist. Mineral catalysts implies that organic synthesis could also occur in deep water where ultraviolet light has been filtered."[13]

Until recently there was no experimental proof for the clay hypothesis. Montmorillonite clay has been shown to have the property to form phosphodiester bonds. A phosphodiester bond is a bond between a two sugar groups and a phosphate group. A phosphodiester bond links two nucleotides together to form DNA and RNA, the nucleotide polymers. These bonds are central to all life on Earth, as they make up the backbone of the strands of DNA. Initial studies established that some montmorillonite clays catalyze condensation of activated mononucleotides to oligomers.[14] Therefore, clay played a role to catalyze nucleotides into DNA. Subsequently the DNA became more complex and organized, through various stages, into a living cell. The Quranic and the medieval Muslim description of the role of clay minerals in the creation process are discussed in the next two chapters.

Another scientific claim, by Manfred Eigen, William Gardiner, Peter Schuster, and Ruthild Winkler-Oswatisch is that the earliest genetic unit was ribonucleic acid (RNA). They assumed that RNA preceded DNA because the function of the complex DNA structure depends upon RNA. Thus, they presumed that "life began on the day when one or more replicating RNA molecules were formed by accident [and that] after a period of competition and evolution, a winner emerged."[15] Eigen's theory posits that RNA molecules by themselves rearranged their structure and somehow directed themselves towards a synthesis of more complex enzymes, which led to the "accidental" development of life.

Maitland Edey, editor of Time-Life Books, and Donald C. Johanson, director of the Institute of Human Origins, critique Eigen's theory. They find it "odd to speak of things like 'selection,' 'competition,' and 'success' taking place among the inert chemicals." In their view,

Eigen's model is weak because "aside from the initial step, none of the next ones have been shown to have happened." Moreover, they point out that Eigen's mathematical calculations are flawed.[16] It may seem odd for Edey and Johanson as well as other materialists to speak of competition between "inert chemicals." However, within the Islamic theory of creation described in this work, we will see how such competitions between all so-called "inert chemicals" and all big or small components of the universe mesh with Allah's process of creation.

According to evolutionists, the primitive cell gradually evolved into plants and animals. The process of evolution occurs due to spontaneous mutations or changes in the genetic material within the cell. Accumulation of these small mutations and the isolation of the organism from its parental species results in speciation. Most evolutionary biologists believe that speciation is a gradual process, with no clear-cut line between different non-interbreeding species. Scientists believe that all life forms are changing and that humans may evolve into another form, the nature of which is not known to us. One of the most important corollaries of the modern theory of evolution is that evolution is non-linear and non-directional. All life is continuing to evolve with humans at a branch of a growing tree, not at the end of the tree. That is, modern apes and humans share a common ancestor. Spontaneous mutation and the idea of the possibility of humankind evolving into another species serve as nothing new within the medieval Islamic concept of creation. These points are discussed later in this book.

In *The Origin of Species*, Charles Darwin proposed that in nature, there is a struggle for survival between organisms, which he calls natural selection and some scientists call survival of the fittest. He described the outcome of this struggle as follows:

> "Owing to this struggle, variations, however slight and from whatever cause proceeding, if they be in any degree profitable to the individuals of a species, in their indefinitely complex relations to other organic beings and to their physical conditions of life, will tend to the preservation of life and will tend to the preservation of such individuals, and will generally be inherited by the offspring. The offspring, also will have better chance of surviving, for, of the

many individuals of any species which are periodically born, but a small number can survive...from the war of nature, from famine and death, the most exalted object which we are capable of conceiving, namely, the production of the higher animal directly follows."[17]

Darwin concludes that "there is grandeur in this view of life, with its general powers having been originally breathed by the Creator into few forms or into one; and that, whilst this planet has gone cycling on according to the fixed laws of gravity, from so simple a beginning endless forms, most beautiful and most wonderful have been, and are being evolved."[18] Later in this book we will see that the idea of evolution was well-accepted by Muslims centuries before Darwin, and that they describe it almost verbatim as Darwin describes here.

G. H. Hardy and W. Weinberg, the fathers of evolutionary mathematics, explained the mechanism of evolution. They state that Mendel's laws of inheritance predict the outcome of single-pair mating but do not predict the proportions of genotypes in a population. A genotype is the genetic constitution of (or the totality of genes possessed by) an organism or a group irrespective of its physical appearance.

The genotype reflects the frequency of alleles in the population. Alleles are the pairs of alternative forms of gene, one from each parent, for a particular Mendelian character—for example, black or white complexion—occupying the same position in a homologous chromosome. Hardy and Weinberg found that both allelic and genotypic frequencies remain unchanged from generation to generation unless their equilibrium is disturbed by external factors such as natural selection, mutation, migration, or genetic drift.[19]

An example of natural selection that disturbs equilibrium involves the change in the frequency of the gene for melanin in our friend *Biston betularia*. Before industrialization in the United Kingdom, the dark-colored peppered moths with melanin genes were fewer in number than light-colored moths. Environmental pollution blackened the barks of trees where the moths rested in the daytime, allowing the dark-colored insects to blend into the darkened barks. The light-colored moths became more obvious prey for the birds. This led increased survival and reproduction of the moths that carried the melanin gene. In the 1950s, when a significant decrease in the industrial pollution occurred, the light-colored moths again began to grow in number. This transformation at the subspecies level is a well-documented example of natural selection caused by external factors within historical time.

Other than natural selection, the migration of organisms to other host populations affects allelic and genotypic frequencies as a result of mating between migrant and host populations. This is known as gene flow, as it causes changes in the gene pools. Such changes take place, for example, in plants when the wind spreads seeds or pollen beyond the bounds of a local plant population.

Still another important factor that alters the allelic frequency is genetic drift. When a change in the proportions of alleles in a gene pool occurs by chance, various outcomes can result. The alleles, which form the gene pools of the next generation, are samples from those of the current generation. Sampling new alleles can be compared to sampling marbles from a bag filled with nine hundred black marbles and one hundred white marbles (90% black and 10% white). If someone picks up ten marbles at random without looking into the bag, there is a chance that he will get no white marbles. Thus, pure chance can cause deviations in the expected frequency of white marbles. Such random changes in the gene pools of a particular population is called genetic drift.

As in the marble sampling case, random selection operates in the propagation of alleles from parent population to offspring population. Human beings have no way of predicting the exact constitution of an individual offspring, but they can predict the gene pool of a certain population if mutation gets excluded. When we look at the fertilization process, organisms form many more gametes than those actually taking part in such fertilization. There is always a chance that an allele will be missing from those that contributed to the genetic constitution of the next generation, primarily due to sampling error (random selection of one gamete out of a few million that fertilize one ovum). Thus, the gene pool changes by chance alone.[20]

As Christian fundamentalist creationists and some Muslims use the words random and/or chance to discredit science and the theory of evolution, we should briefly clarify the scientific meaning of these words. Gould defines random thusly:

> "A phenomenon governed by chance yields maximal simplicity, order, and predictability—at least in the long run....If we flip a penny or throw a pair of dice, once a second for days on end, we achieve a rigidly predictable distribution of outcomes. We can predict the margins of departure from 50-50 in the coins or percentage of sevens for our dice, based on the total number of throws. When the number of tosses becomes quite large, we can give estimates and ranges of error for the frequencies and lengths of runs—heads or sevens in a row—all based on the simplest of mathematical formulas from the theory of probability. (Of course, we cannot predict the outcome of any particular trial or know when a run will occur, as any casual gambler should—but so few do—know)."[21]

To discredit the theory of evolution, "scientific" creationists argue that the human eye defies evolution. They posit that the complex system of the lens, iris, photoreceptors, and special nerves of the eye cannot be explained by the randomness of evolution. They say that the odds of an eye evolving are equivalent to a monkey typing Shakespeare's *Hamlet* which could not happen by random chance.

Evolutionists counter the argument of so-called scientific creationists by asserting that natural selection is not random, but a trial and error process that "preserves gains and eradicates mistakes." They support their claim with examples:

> "The eye evolved from a single, light-sensitive cell into the complex eye of today through hundreds if not thousands of intermediate steps, many of which still exist in nature. In order for the monkey to type the first 13 letters of Hamlet's soliloquy by chance, it would take 26 to the power of 13 number of trials for success. .if each correct letter is preserved and each incorrect letter eradicated, the process operates much faster. How much faster? Richard Hardison constructed a computer program in which letters were selected for

or against, and it took an average of only 335.2 trials to produce the sequence of letters TOBEORNOTTOBE. This takes the computer less than 90 seconds. The entire play (of *Hamlet*) can be done in about 4.5 days."[22]

Moreover, we have evidence that the eye may have evolved independently many times during the history of life. It progressed from a simple eye spot made of light-sensitive cells, as in the flat worm, to individual photosensitive units, as those in insects with light-focusing lenses, to the eventual form of an eye with a single lens focusing images onto a retina. In man and other vertebrates, the retina consists not only of photoreceptor cells but also of neurons that analyze the visual images. Through gradual evolutionary steps, different kinds of eyes have evolved, from simple light-sensitive organs to highly complex systems of vision.

The fact that evolution has taken place is no longer a disputed issue among the majority of biologists. However, as D. J. Futuyma, professor at State University at New York, tells us, theoretical explanations of the causes for evolutionary change remain controversial.[23] Two predominant theories explain the patterns of evolution: the neo-Darwinian evolutionary synthesis, which Ernest Mayr of Harvard University calls phyletic gradualism, and Gould's punctuated equilibria.

The modern evolutionary synthetic theory holds that small mutations take place in an organism, resulting in small variations—in size, strength, shape, behavior, intelligence, endurance, and so on—among the new generations. Ecological factors include natural selection screening out some combinations of mutant genes to form subspecies. Isolating mechanisms, such as geographic barriers, habitat preference, seasonal breeding, behavioral (ethological) differences, etc. prevent the subspecies (incipient species) from merging with the general populations from which they originated. Thus, natural selection aided by the isolating mechanisms produces good species.[24] In the words of Ernest Mayr, the evolutionary synthetic theory or "gradual evolution . . . [is] explained in terms of small genetic changes (mutation) and recombination, and the ordering of this genetic variation by natural selection."[25]

The implication of classical Darwinism and neo-Darwinism is that the fossil record should reflect slowly changing forms—that the intermediate forms of life should be preserved as fossils and that we should be able to unearth them. Darwin wrote in *The Origin of Species:* "I look at the geological record as a history of the world imperfectly kept, and written in a changing dialect; of this history, we possess the last volume alone, relating only to two or three countries. Of this volume, only here and there a short chapter has been preserved; and of each page, only here and there few lines."[26]

Many more fossils have been discovered over the hundred years since Darwin's death, of course. We are no longer limited to the most recent volume, and some of the volumes are now far more complete. Nevertheless, fewer examples of gradual change have been found than might have been expected. Therefore, many evolutionary biologists have questioned the modern evolutionary synthesis theory. Steven Stanley, a paleontologist at Johns Hopkins University, points out that there is no record of phyletic evolution to validate gradualism.[27] Similarly, David M. Raup, geologist and curator of the Field Museum of Natural History in Chicago, maintains that "[I]nstead of gradual unfolding of life, what geologists of the present day actually find is a highly uneven or jerky record; that is, species appear in the sequence very suddenly, show little or no change during their existence in the record, then abruptly go out of the record."[28]

From Darwin's time until recently, discrepancies in the model of phyletic change in the fossil record have been ascribed to imperfections in the fossil record itself. Other causes are postulated to be erosions, lack of proper conditions for fossilization, or simply not looking at the right place. Nils Eldredge and Stephen Jay Gould have a different explanation for the morphological invariance of species through time and the unconnected intermediates in the fossil record. The theory of punctuated equilibrium by Eldredge and Gould explains the rarity, not absence, of transitional forms. Gould posits punctuated equilibria as a sound explanation for the inconsistencies of the fossil record. He dismisses the idea of gradual evolution and affirms the idea of evolution that is more comparable to "climbing a flight of stairs (punctuation and stasis) than rolling up an inclined plane."[29] Gould states:

"New species usually arise, not by the slow and steady transformation of entire ancestral populations, but by splitting off of small
populations from an unaltered ancestral stock. The frequency and
speed of such speciation is among the hottest topics in evolutionary theory today, but I think that most of my colleagues would
advocate ranges of hundreds or thousands of years for the origin
of most species by splitting."[30]

Some Christian and Muslim anti-evolutionists distort the theory
of punctuated equilibrium by propagating the idea that this theory
does not accept the existence of any transitional species. But punc-
tualism clearly postulates the existence of morphological intermediates as small populations. Punctualists claim that evolution is
continuous but jerky, with established lineages.[31] A species remains
static without any morphological change. Then, a small peripheral
isolated population of the species changes rapidly into a new species as a result of mutation in the regulatory gene that controls the
expression of the structural gene.[32] The new species migrates into
the parent species. Gould and Eldredge have not claimed their explanation of gaps in the fossil record to be "a new discovery, but a
novel interpretation for the oldest and most robust of paleontological observations: the geologically instantaneous origination and
subsequent stability (often for millions of years) of paleontological
'morphospecies'."[33]

For the anti-evolutionists among the Christians and Muslims, the
phrase "geologically instantaneous" has become music for their ears.
They proudly proclaim that Gould's descriptive phrase is the confirmation of their belief that God created all species abruptly from the
thin air like a magician! It appears that they have a selective deaf-
ness. They hear only what they want to hear. Gould has pointed out
many times that the instantaneous speciation is an artifact of the
compression of time that takes place each time a new layer is formed
in the fossil record. The geological moments are equal to millions of
years in the human time frame.

The study of life in the geological past is called paleontology.
Sedimentary rocks are composed of sediments that gradually settled
in cumulative layers at the bottom of oceans and large lakes, carrying

with them the remains of plants and animals. The deepest strata of these rocks are the oldest, and the shallowest are the most recent.

Paleontologists have recovered and scrutinized the fossil remains of thousands of organisms that lived in the past from these sedimentary rocks, and they have arranged them according to the geological time scale. Fossil records show that many kinds of extinct organisms were different in form from those living now. They reveal successions of organisms, with simpler life forms appearing early and complex forms appearing later

The stratigraphic dating as described in Table 5-1 is reliable but not absolute. It shows only the chronological order for the appearance of organisms, but does not tell the exact time of their appearance within the chronological age of the earth. Absolute dating is determined by radiometry, which uses the rate of decay of radioactive elements in the geological deposits and fossils to calculate their age. Radioactive dating has shown that the earth was formed 4.5 billion years ago. The study of fossils clearly indicates that not all forms of life originated simultaneously, nor did they appear in six days.

The earliest fossils of microorganisms resembling bacteria and blue algae are three to five billion years old. The oldest animal fossils are of small wormlike creatures. They are seven hundred million years old. Animals with backbones, known as vertebrates, appeared as primitive fishlike organisms four hundred million years ago and gradually evolved into modern fish. Later, amphibians and birds appear in the fossil record. Finally, mammals appear in fossils of about two hundred million years old. (Mammals are a class of warm-blooded vertebrates who secrete milk to feed their young). The most primitive, single-celled organisms are found in the oldest and deepest rocks, while chronologically younger rocks contain progressively more complex forms of life.[34] Such a distribution and layering suggests evolution for many agnostics and creation by modification for many God-conscious people of all faiths.

Comparative anatomy is the study of inherited similarities in bone structure and soft tissues among various organisms. Even without fossil evidence, we are able to infer that organisms have evolved by adapting structures of other species that still survive today. The skeletons of bats, whales, humans, and horses are remarkably similar,

Table 5-1. Organisms in Geological Time Scale

Era	Millions of years before present	Biological Events
CEMOZOIC	0.3	Modern man
(AGE OF MAMMALS)	1.0	Homo erectus
	2.0	Homo habilis
	5.0	Australopithecus afarensis
	25.0	Ramapithecus
	37.5	Old world monkey
	55.0	New world monkey
	75.0	First placental mammals
MESOZOIC	135	Climax of reptiles,
(AGE OF REPTILES)	185	Reptiles dominant; First bird; first mammals. First flowering plants.
	225	First dinosaurs; first appearance of frogs, crocodiles, mammals, flying reptiles; conifers dominant.
PALEOZOIC	275	Widespread marine life extinction; mammal-like reptiles flourish.
	310	First reptiles appear; Giant insects on land; amphibians thrive in sea
	350	Echinoderms abundant in seas; terrestrial amphibians rare.
	395	First bony fishes; sharks.
	435	Earliest known land plant; Fishes become important.
	480	Earliest known fishes.
	600	Organisms with hard parts appear; fragmentary vertebrates appear.
PRECAMBRIAN	3000–4800	Soft bodied primitive life protozoans; molluscs; annelids; arthropods; others of uncertain affinities; bacteria.

even though they live in different environments and maintain very different lifestyles. Homologous versions of different bones are found not only in the limbs of the various animals but throughout the remainder of their bodies as well. The various homologous bones enable humans to write, bats to fly, whales to swim, and squirrels to carry and hide their food for the coming winter. The concept that each individual species emerged separately from others is not congruent with the evidence of similar bone structures found across different types of animals. Similarities of skeletal anatomies can be explained as structures inherited from common ancestors modified to adapt to the different environments and ways of life that each species had to deal with.[35] (Figure 5-3.)

Further evidence of common ancestral species can be gained from studying the development of animal embryos. The early embryos of vertebrates such as fish, birds, lizards, and mammals (including humans) are strikingly similar. The fact that most animals develop gill slits below their heads suggests that these animal embryos may have descended from ancestral species that lived in the water, although many embryologists believe that these are not gills but mere folds (flexures).

Scientists have found that across different species, structures analogous to the bones of the forehand all develop from the same region of the embryo. Moreover, vertebrate embryos initially have tails. In certain species (including humans), tails disappears as embryos mature, except in some extremely rare cases in which human babies are born with tails. This common developmental pattern across various species of animals and humans reflects our evolutionary kinship within the animal kingdom.[36] (Figure 5-4.)

Vestigial organs are structures that are greatly reduced, often without any function in the body. Such vestiges of the embryonic rudiments are common in all sorts of animals, including human beings. For example, in mammals that have a coarse diet with considerable amounts of cellulose, a part of the large intestine called the caecum forms a large sac in which mixtures of food and enzymes can react for a long period of time; a constricted appendix is absent altogether. But in human beings the caecum is modified into an appendix. The nictitating membrane of the eyes is another example. All tetrapods

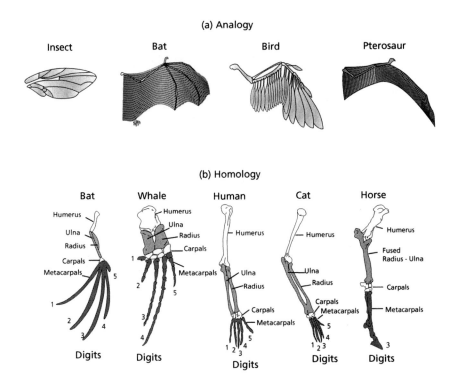

FIGURE 5-3. ANALOGY AND HOMOLOGY. (a) Analogy: Wings of a fly, a bird, and a bat. The wings of a fly, a bird, and a bat are anologous structures that are developed from different materials. Evolution of each was indepedant. (b) Homology: limbs of a bat, whale, human, and horse. Bones with the same intensity of shading are homologous throughout, even though they are modified in size and in shape by reduction or by fusion of bones. Some materials have been changed to meet the needs of different animals.

(four-footed animals) have a nictitating membrane in the inner corner of the eye. In most vertebrates, this membrane sweeps clear across the eyeball to cleanse it. In birds, this membrane is particularly well developed. In horses and other mammals, it is well developed and fully functional. But in humans and some other mammals, it forms a mere fold at the inner corner of the eye and has no function. The human external ear muscles present a similar situation. Many mammals move the external ear freely in order to detect sounds efficiently. The complete muscular apparatus for these movements is present in man, but it has no function.[37] (Figure 5-5.)

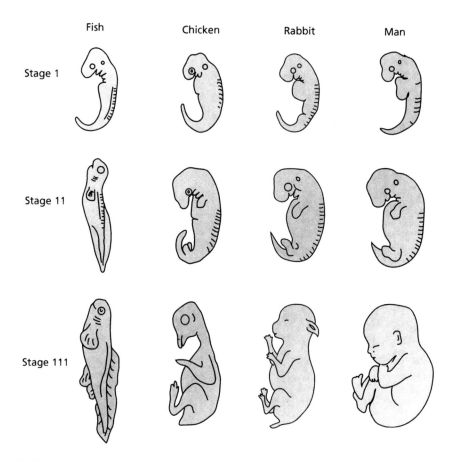

FIGURE 5-4. EMBRYOLOGICAL EVIDENCE FOR EVOLUTION. Three parallel stages in the development of four vetbrates; fish, chicken, rabbit, and man, show how the early embryonic stages resemble one another. This suggests common ancestry and creation by modification (Islamic concept) or descent modification (Secular concept).

As previously mentioned, DNA is the cornerstone of all life on earth. It is made out of phosphate, sugar, and four bases—adenine (A), cytosine (C), guanine (G), and thymine (T). (Figure 5-6.)

These bases can be thought of as a four-letter alphabet or code. Just as words in our language differ according to the sequence of letters of the alphabet, the sequence and number of the bases (A, C, G, and T) in the DNA determine the uniqueness of a gene; and combinations and permutations of genes in the chromosome decide the

Figure 5-5. Few of the vestigial structures in human beings.

nature and shape of life forms on earth. The study of chromosomes quite convincingly points out that all organisms, from bacterium to human, are made out of the same substance. In terms of molecular composition, they differ only in relation to the number and sequence of the four bases in each gene and the combinations of genes in each chromosome, as well as in relation to the number of chromosomes in each cell. This unity in diversity suggests the genetic continuity and common ancestry of all organisms.

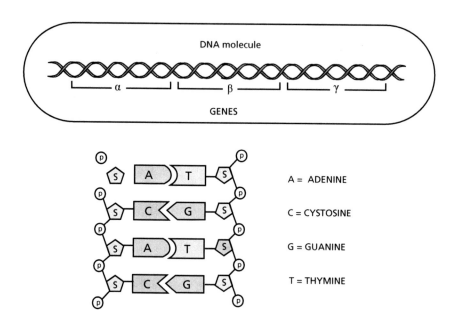

Figure 5-6. STRUCTURE OF DNA

The evidence of evolution presented above serves as a fraction of examples too numerous to include here. Unquestionably we can conclude that biologists have posited various theories to explain (1) the phenomenon called life, with all its peculiarities and varieties; (2) the closeness and remoteness in time of various life forms and the interrelationships between them; (3) the geological appearance of less complex life forms during the early stages of the earth as well as the appearance of graduated complexity in the organisms during the long history of the earth; and (4) the interrelationship between organisms and the environments in which they live.

Universally, biologists do not agree on all postulates, but the overwhelming majority agrees that the different forms of life did not originate simultaneously. They also agree that in the early biography of life, organisms were much less complex, but over a period of 3.5 billion years on the earth increasingly complex life forms emerged. Similarly, almost all biologists concur that organisms can be traced to a single family, based upon the overwhelming evidence of later organisms having evolved from their earlier ancestors.

Evolutionists readily admit that details of such origination and the mechanisms of the transformation of simple organisms into multitudes of species still remain a matter of debate. Unfortunately, American Christian evangelical fundamentalist organizations such as the Institute for Creation Research present a distorted version of such admissions, claiming that the evolutionary puzzles yet to be solved somehow negate the facts that scientists do know. Sadly, some contemporary Muslims also are drawn into the ICR's distorted presentation of scientists' genuine disagreements among themselves.

CHAPTER 6

Pre-Darwinian Muslims and the Theory of Evolution

T HROUGHOUT HISTORY, Muslims had long recognized that humans have an ancestral link to apes, and they taught this fact in schools which, at the time, were extensions of mosques. Nonetheless, Judeo-Christian individuals in the West were shocked when they first read Charles Darwin's *The Origin of Species* (1859). Their immediate response was generally similar to that of the wife of the Bishop of Worcester, who after hearing that humans evolved from monkeys was reported to have exclaimed, "Descended from apes! My dear, let us hope it is not so; but if it is, that it does not become generally known."[1]

The theory of evolution described by Darwin's in *The Origin of Species* had once been widely known, however; his book was only a confirmation of a centuries-old Muslim theory of evolution through data that he had collected in his travels. John William Draper (1812-1883), a prominent scientist, evolutionist, and professor of chemistry at New York University, was a contemporary of Charles Darwin. In 1838 he made the first photographic portrait of the moon through a chemical process.[2] Among his many works are his *History of the*

Intellectual Development of Europe and *The History of the Conflict between Religion and Science.*

Six months after the publication of Darwin's book, Professor Draper presented a paper at the meeting for the British Association for Advancement of Science entitled "The Intellectual Development of Europe Considered with Respect to the Views of Mr. Darwin." During the discussion of this paper, Bishop Wilberforce of Oxford contemptuously inquired of Thomas Huxley, an eminent scientist and advocate of the theory of evolution, whether Huxley claimed his descent from monkeys "through his grandfather or grandmother."[3 & 4]

Draper pointed out in *The History of the Conflict between Religion and Science* that the Fathers of the Church insisted upon a more recent origin of Creation. He explained such insistence as a justifiable response to claims that God had neglected the human race as a whole in favor of "the few who were living in the closing ages of the world" for salvation. Draper explained that Christian teachings of the story of the perfect world of Adam served as a necessary premise for the fall of humanity, which in turn was a prerequisite for "the plan of salvation." In Draper's view:

> "[Christian] (t)heological authorities were therefore constrained to look with disfavor on any attempt to carry back the origin of the earth to an epoch indefinitely remote, and on the *Muhammadan theory of evolution* [italics mine] which declared that human beings developed over a long period of time from lower forms of life to their present condition."[5]

Draper, an independent western source, knew that Darwin's theory of evolution was actually the Muslim theory of evolution. One finds it difficult to believe that European scientists, who were neighbors to the Muslims, were unaware of their theory of evolution when an American scientist, a contemporary of Darwin, knew that the theory originated among the Muslims. Is it possible that Europeans, who followed Roger Bacon's advice to "learn Arabic and Arabic science for progress,"[6] did not know about the Muslim theory of evolution?

In our own time, there is a sharp rejection of evolutionary studies. The free press in the West, especially that of the United States, avoids the mention of the word "evolution" in some public school textbooks, fearing condemnation from some Western religious establishments. For example, *Man: A Course of Study (MACOS)*, begun in 1963 by the National Science Foundation, was finally published in 1970 by the Education Development Center as an introduction to evolution for elementary school students. "No commercial publisher would touch the project because religious groups would not endorse the teaching of this type of material," until a small foundation agreed to publish it. By 1974, many school districts in forty-seven states had adopted *MACOS*, but in 1975, when organized opposition began to assert itself, this sales rate plummeted seventy percent. "The National Science Foundation, which had provided $4.8 million to develop MACOS, suddenly was attacked in Congress. The House of Representatives passed the Bauman Amendment in 1975, giving Congress direct supervision and veto power over every single NSF research grant, [and] Nineteen eighty-one saw a drastic cut in federal support for social science research, and science education was virtually eliminated from the federal budget."[7]

Contrary to the current opposition to teaching evolution in American public schools, the doctrine of the gradual development of life, humankind, and other life forms was part of the curriculum in Muslim schools centuries before Darwin. Draper discredited the Western myth that Lamarck and Darwin were originators of the theory of evolution and declared that the Muslim theory was more advanced than that of Darwin.

> "Sometimes, not without surprise, we meet with ideas with which we flatter ourselves with having originated in our own times. *Thus our modern doctrine of evolution and development were taught in their* [Muslim] *schools* [italics mine]. In fact they carried them much farther than We are disposed to do, extending them even to inorganic or mineral things."[8]

Most Western evolutionists agree that Darwin's grandfather, Erasmus Darwin (1731–1802), influenced his grandson's interest in evolution.[9] Little is written, however, about influences on their thoughts on the subject. Erasmus Darwin was a philosopher, a botanist, a poet, a physician, and the founding member of a philosophical society called "the Lunatics."[10] In Erasmus's day and long before, almost all philosophical and scientific books were translations or retellings of Arabic books from Muslim scientists and philosophers.[11] Erasmus Darwin, being a philosopher and physician, most likely had discussed Muslim philosophical and scientific work during the meeting of "the Lunatics." I believe, therefore, that the Darwins learned evolutionary biology from Muslim scientists, which is confirmed by the history of medical education in Europe. Professor Draper describes the state of Islamic and European medicine before the Enlightenment of the West as follows:

> "Saracens commenced the application of chemistry, both to the theory and practice of medicine, in the explanation of the functions of the human body and in the cure of diseases. Nor was their surgery behind their medicine...How different was all this from the state of things in Europe: the Christian peasants, fever-stricken or overtaken by accident, hied to the nearest saint-shrine and expected a miracle; the Spanish Moor relied on the prescription or lancet of his physician, or the bandage and the knife of his surgeon."[12]

Will Durant, the American historian, informs us that the Manual for Oculists, written by the great Muslim occultist Ali ibn-Isa, was "used as text in Europe till the eighteenth century." Durant also stresses the importance of the works of Abu Bekr Muhammad al-Razi (844–926), better known in Europe as Rhazes, who was one of many Muslim scientists and healers. According to Durant, Al-Razi's twenty-volume book, *Kitab al-Hawi* (The Comprehensive Book), which covered all branches of medicine and was translated into Latin, was "probably a highly respected and frequently used medical textbook in the white world for several centuries" and was one of nine books used at the University of Paris in 1395. Another famous figure whose philosophical and Scientific contributions are illuminated by Durant is Abu Ali al-Husein ibn Sina (Avicenna) (980–1037), whose books were encyclopedias of knowledge that included studies in mathematics, physics, physiology, hygiene, therapy, pharmacology, philosophy, metaphysics, theology, economics, politics, and music. His books were taught as main texts in "the universities of Montpellier and Louvain till the middle of the seventeenth century."[13] Two other Muslim physicians who influenced Europe and European medicine were Abu Bekr ibn Tufail (Abubacer) (1107–1185) and his student Abu al-Walid Muhammad ibn Rushd (Averroes) (1126–1298). Ibn Rushd wrote an encyclopedia of medicine (*Kitab al-Kulliyat fi-l-tibb*) that was translated into Latin and taught in Christian universities. All Muslim physicians were evolutionists, and Western historians have acknowledged the fact that books of medicine written by physicians of the Golden Age of Islam served as the standard textbooks used in all medical schools in Europe until the eighteenth century. Therefore, Charles Darwin's (1809-1882), grandfather and father, Erasmus and Robert Darwin, both physicians, were undeniably influenced by the above-mentioned textbooks.

Lastly, the first Latin translation of Abu Bakr ibn Tufail's *The Story of Hai bin Yaqzan (The Journey of the Soul)* by Edward Pocock, Jr. was published in Oxford in 1671. Several editions of this work appeared in the years from 1671 to 1700. Then, in 1708, Simon Okley published the first English translation, and Dutch, German, French translations were made in eighteenth and nineteenth century.[14] The publication of many editions and different translations of this book in England and other parts of Europe in the sixteenth to eighteenth centuries suggest that it was a very popular book; the probability is great, therefore, that Charles Darwin, his father, and his grandfather, all read it. Moreover, Abu Bakr ibn Tufail's *The Story of Hai bin Yaqzan (The Journey of the Soul)* is an allegorical tale of the theory of evolution of life and human society. Erasmus Darwin's *The Temple of Nature*[15] is mostly a poetic rendition in English of Tufail's celebrated work.

Clearly, Erasmus and Robert Darwin most likely learned about the Muslim theory of evolution from the translations of Muslim books. A slight chance exists that these erudite men did not know that the authors of their textbooks of medicine and philosophical essays were evolutionists. In that case, such ignorance can only be explained as an unlikely historical oddity. If, as is far more likely, they did learn about evolution from Muslim scholars, their failure to credit their sources appears to be an intentional plan to obscure the pioneering Islamic contributions to the study of evolution. The evidence presented above makes the second premise more likely.

Muslims nowadays, like many non-Muslims in the West, resist the idea of evolution. Curiously, these are the same Muslims who boast about the scientific contributions made by their ancestors. They gleefully ask others whether they have heard about Al-Biruni, who, like Bacon, wrote in his *Vestiges of the Past (Atharul- Baqiya)*, "We must clear our minds . . . from all causes that blind people to the truth—old custom, party, spirit, personal rivalry or passion, the desire for influence."[16] Muslims never fail to remind us that Ibn Sina's *Canon of Medicine (Qanun-fi-l-Tibb)* was the chief textbook of medicine in European medical schools until the seventeenth century,[17] and continue to praise al-Haitham's contributions to science. Yet if they were told that these great scientists, of whom they are rightfully so proud, were evolutionists, they would be amazed and annoyed.

Most Muslims in the world believe that Adam and Eve were created in Paradise. They have been indoctrinated with the Judeo-Christian belief that God created Adam, then created Eve from Adam's rib. Jews and Christians who accepted Islam early in its history imported this theological doctrine into Islam. We shall discuss later how this Judeo-Christian story of the *ex nihilo* creation of Adam and subsequent creation of Eve was transplanted into Islamic faith. First, however, a review of early Muslim scholarly works will show that these scholars were evolutionists and did not find any conflict between their faith and the theory of evolution.

Ibn Khaldun, the most famous Muslim historiographer and social scientist, who wrote his *Muqaddimah* [An Introduction to History] over 400 years before Charles Darwin, states:

> "One should then look at the world of creation. It started out from the minerals and progressed, in an ingenious, gradual manner to plants and animals. The last stage of minerals is connected with the first stage of plants, such as herbs, and seedless plants. The

last stage of plants such as palms and vines is connected with the first stage of animals, such as snails and shellfish which have only the power to touch. The word 'connection' with regard to these created things means that the last stage of each group is fully prepared to become the first stage of the next group. The animal world then widens, its species become numerous, and, in a gradual process of creation, it finally leads to man, who is able to think and reflect. The higher stage of man is reached from the world of monkeys, in which both sagacity and perception are found, but which has not reached the stage of actual reflection and thinking. At this point we come to the first stage of man (after the world of monkeys). This is as far as our (physical) observation extends."[18]

The reference in the above passage to the "first stage of man" clearly show that Ibn Khaldun knew that other hominoid species, more advanced than the monkey but not equal to modern man, were created before modern man emerged. Moreover, he states that he arrived at his conclusion from physical observation. As we have seen in the section on uniformitarianism, Ibn Khaldun believed that the human races originated as a result of natural causes. According to the *Muqaddimah*, Ibn Khaldun and other Muslims believed that a series of transmutations of one species into another over a long period of time resulted in the gradual evolution of life, from primitive organisms into a bush with numerous branches. Thus, life forms are not independently created but are evolutionary products from ancestral species.

Ibn Khaldun concludes his view on the origin of the races as follows: "Physical circumstances and the environment are subject to changes that affect later generations; they do not necessarily remain unchanged."[19] According to him, species are not fixed but subject to changes with a changing environment. Even human species are subject to change. He believed that the physical characteristics of organisms are determined by their "essence." He maintained that active nature (*kiyan*) "has the ability to generate substances and change essences"[20] and that earthly existence is a continuum of transformations of essences that occur in stages in a natural order of ascent and descent. He describes the transformation of species into other species as a result of modification of essence (gene) by nature:

> "The essences at the end of each particular stage of the worlds are by nature prepared to be transformed into the essence adjacent to them. This is the case with the simple material elements; it is the case with the palms and vines (which constitute) the last stage of plants, in their relation to snails and shellfish, (which constitute) the (lowest) stage of animals. It is also the case with monkeys, creatures combining in themselves cleverness and perception, in their relation to man, the being who has the ability to think and to reflect. The preparedness (for transformation) that exists on either side, at each stage of the worlds, is meant when (we speak about) their connection."[21]

If he were writing today, Ibn Khaldun would replace the word *essences* with the term "genes" or "DNA." He would be saying: "Nature prepares the genotype (essence) of species to be transformed into the genotype (essence) of the adjacent species." Similarly, instead of saying "transformed into the next stage by nature," he would say: "Gradual evolution can be explained in terms of small genetic changes (mutation) in the species and by the ordering of this genetic variation by natural selection."

Abu Bakr Muhammed ibn-Arabi (1165–1240), one of foremost interpreters of the Quran, was born in Spain. By the time he was twenty-five years old, he was famous throughout the land as a brilliant scholar, a great writer of elegant poetry, and a Sufi of the first order. Between 1193 and 1198, he wrote *Uqlatu'l-Mustawfiz* [Controller of The Wanderer]. When his opinions were severely attacked in the West, he felt his life was becoming intolerable, so in the year 1200 he left Spain for Mecca, after which he traveled extensively until he settled in Damascus. His description of the gradual evolution of species from one origin is similar to that of the other Muslim scholars mentioned in this chapter.

"On they rolled to perfection: Thus the meaner world was born. Mineral passed to vegetable life, out of which animal life was born.[22]

"Then creation continued on earth, minerals, then vegetations, then animals, and then man. God made the last of every one of these kinds. The last of the minerals and the first of the vegetations is the truffle. The last of the vegetations and the first of the animals is the date palm. The last of the animals and the first of mankind is the monkey.[23]

"Since the perfect man is in the perfect form, he deserves the vicegerentship and deputyship of God in the Universe. Here we shall explain the evolution of this vicegerent, his position and form as they are. We do not mean Man only as animal, but, on the other hand, as Man and vicegerent. On account of his human quality and vicegerentship man deserves his perfect form. Every man is not a vicegerent. In our opinion, the animal man is not a vicegerent . . . . This is the intended Perfect Man. Others are animal men. The relationship of animal man to the Perfect Man is that of the ape to the animal man.[24]

"As for the animal man, he is not a man essentially. His case is like that of animals. But he is distinguished from another through differentiae peculiar to every one of the animals.[25]

"The goal of all this was man coming in perfect form. When the field was thus prepared, Man came in the nicest form.[26]

"When God desired the perfection of human evolution, He collected and bestowed on Man all realities of the Universe and illuminated him with all His names.[27]

"When this comprehensive name became capable of two aspects by itself, it became fit for vicegerentship and organization and gradation of the Universe. If Man does not reach the stage of perfection, he is an animal whose appearance resembles the external appearance of man. Here we are concerned with Perfect Man. The first of human species whom God made was the Perfect Man. He was Adam (peace be upon him). Thus God demonstrated the stages of perfection for the species. He who attains to it is the man who attains perfection, and he who goes down from that stage is one who possesses the human quality in proportion to where he is."[28]

Ibn Arabi explains the concept of the "perfect man" as being God's intended vicegerent on earth. In order for a human being to become God's deputy on earth, he had to evolve into a perfect form. Ibn Arabi makes a distinction between a "perfect man" and an inferior form which he calls "animal man." The latter is "not a man" because he exhibits qualities peculiar to animals. His relationship to a human being is like "that of the ape to the animal man." In Ibn Arabi's view, if man does not reach the stage of perfection, he is an animal whose appearance resembles the external appearance of man. Thus, it was God's intention for human beings to evolve into their perfect form[s] so they would be suitable vicegerents of the Universe.

The above ideas from Ibn Arabi demonstrate that the Muslims knew about the existence of hominoid species after the advent of the ape and before the emergence of modern Homo sapiens. Not only did Ibn Arabi classify the hominid as an "animal man," he made the distinction between that species and the human being he called "perfect man."

Gould asked "why the Lord saw fit to make so many kinds of hominids, and why some of his later productions, Homo-erectus in particular, look so much more human than the earlier models?"[29] Ibn Arabi anticipated these questions long before Gould or Darwin. Fossils of the animal man to which Ibn Arabi referred were only recently unearthed and classified as Australopithecus, Homo erectus, Homo ergaster, Neanderthal man, and Cro-Magnon man. An artist's reconstruction of human evolution based on modern fossil findings (Figure 6-1) fits well with Ibn Arabi's description of the origin of humans from earlier species. It is possible that Ibn Arabi and his contemporaries arrived at the conclusion that man evolved from "animal man" to Human not by pure speculation, but by scientific methods.

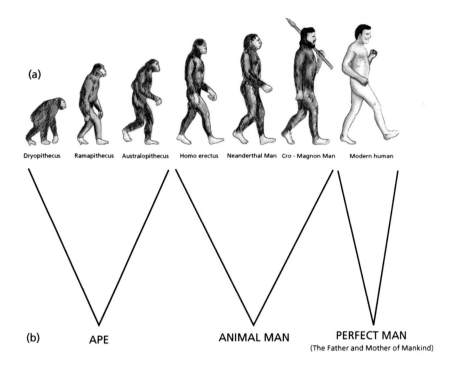

FIGURE 6-1. (a) An artistic reconstruction of the evolution of man and the classification of species based on fossil evidence (b) The pre- Darwin Muslim classification of man and his ancestors

For example, Al-Biruni came to the conclusion that there were hominids of different statures by studying the size of their bones, which were buried in the caves of the mountains of Median where they lived. Al-Biruni describes these bones as those of large races of men "with bones as large as camel-bones and even larger."[30]

Predating Ibn Arabi in the tenth century, Ikhwan al-Safa (the Brethren of Purity) wrote about the gradual development of life. They maintained that species developed successively from minerals to plants to animals and finally into humans, that imperfect species preceded the more perfect ones, and that God created them for humankind's sake. They pointed out that "by the grace of God" nature provided each species with the necessary organs to insure its survival and secure the completion of its perfect form. These conclusions are derived from the following quotations from the *Rasa'il of Ikhwan al-safa* (The Epistles of Ikhwan al-Safa).

> "The first stage of the plant kingdom is connected with the last stage of minerals and the highest stage of the plant kingdom with the first stage of animal.... Also we would like to show in this letter that the highest stage of the animal kingdom is connected with the first stage of the human.[31] Be it known to you! The imperfect animals preceded the most perfect animals in time and in the process of creation.[32] Be it known to you, brothers! The animals were created for man's sake. (And) everything that is created for the sake of something else will precede the beneficiary.[33] By the grace of God's wisdom and care, animals were bestowed with organs, joints, vessels, nerves, membranes, and chambers according to their needs for benefiting themselves or to avoid injury, so that it can survive and will be completed and perfected to reach the highest stage."[34]

The celebrated mystic poet Jalaluddin Rumi (1207–1273) was born in Balkh, Afghanistan. His father was the son of Bahu'u'ddin Walad, a well-known professor of theology and a student of Ibn Arabi. Rumi's celebrated philosophical treatise, *Masnawi-ye Ma'nawi* (or *Mathavi* in Persian, Rhyming Couplets of Deep Spiritual Meaning), has been linked to the writings of Ibn Arabi. Therefore, Ibn Arabi's thoughts may have influenced Rumi's views on the creation of man.[35] In his treatise Rumi echoed the same perspectives on the origin of man and life as that preached by other Muslim scholars of his time and before. In the *Masnavi*, he wrote:

> "Hundreds of thousands of years I was flying (to and fro) involuntarily, the notes in the air.[36] He came first to the inorganic realm and from there stepped over to the vegetable kingdom. Living long as a plant, he has no memory of his struggles in the organic realm. Similarly rising from the plant to the animal life he forgets his plant life, retaining only an attraction for it which he feels especially in the spring, ignorant of the secret and cause of his attraction like the infant at the breast who knows not why he is attracted to the

mother....Then the creator draws him from animality to humanity. So he went from realm to realm until he became rational, wise,
and strong. As he has forgotten his former types of reason (every
stage being governed by a particular type of reason) so he shall
pass beyond his present reason. When he gets rid of this coveted
intellect, he shall see a thousand other types of reason.[37] I died
from the inorganic realm and became a plant; then I died from
plant life and became an animal. Dying from animality I became a
man, so why should I be afraid of anything less through another
death? In the next step I shall die from humanity to develop wings
like angels. Then again I shall sacrifice my angelic self and become
that which cannot enter imagination. Then I become non-existent
when the divine organ strikes the note, 'We are to return unto
Him'."[38]

According to Rumi, the so-called evolution and natural selection
were acts of God, who "draws him from animality to humanity."
Further, Rumi envisions creation as a journey through which a human being necessarily experiences the passage from a physical to a
metaphysical existence, thereby returning to God who is the ultimate
source of creation. Islam's greatest geologist, Al-Biruni (973–1048),
believed in the gradual development of man and other organisms
and in natural selection through the struggle for the survival of the
fittest. He wrote about his beliefs regarding the progressive development of humankind in his book, Kitab al-Jamahir.

"Man reached his maximum degree of perfection compared to
other animals below him (in the evolutionary ladder). He ascended
to the present state from other kinds of beings such as dog-like,
bear-like, ape-like, etc. Then finally he became man."[39]

Abu Bakr Ibn Tufail (1095–1138) (Latin name: Abu Bazer), a poet
and an eminent physician, was born in Guuadix, Spain and died in
Morocco. He was known to have influenced Jewish and Christian
thinkers and was the teacher of the great philosopher Ibn Rushd
(Latin name: Averroes). Although he wrote treatises on medicine
and philosophy, Ibn Tufail is remembered for his famous work, *The*

Story of Hai bin Yaqzan (The Journey of the Soul).[40] The book is a fictional biography that deals with the idea of evolution of life and human society. Simon Ockley undertook the English translation in 1708, about a century before Darwin was born. H. F. Osborn summarizes the story as follows:

> "There happens to be under the equator an island, where Man comes into the world without father and mother; by spontaneous generation he arises, directly in the form of boy, from the earth, while the spirit, which, like the sunshine, emanated from God, unites with the body, growing out of a soft, unformed mass. Without any intelligent surroundings, and without education, this 'Nature-man' through simple observation of the outer world, and through the combination of various appearances, rises to the knowledge of the world and of the Godhead. First he perceives the individuals, then he recognizes the various species as independent forms; but as he compares the varieties and species with each other, he comes to the conclusion that they all sprung from a single animal spirit, and at the same time that the entire animal race forms a single whole. He makes the same discovery among the plants; finally he sees the animal and plant forms in their unity, and discovers that among all their differences they have sensitiveness and feeling in common; from that he concludes that animals and plants are one and the same."[41]

Osborn points out that the boy compares various species of plants and animals and concludes that they sprang from one common origin. Those who wish to preserve science from the corruption of religion echo the same sentiments. Robert J. Shadewald, the former editor of the *National Center for Science Education Report*, asserts that "conventional science offers the same simple answer—all animals have descended from a common ancestry."[42]

Eight hundred years before Darwin, Ibn Miskawayah (930–1030), a Persian Muslim scientist, philosopher, and historian, was interested in alchemy and poetry and left a legacy of twenty works that served as models for later generations of Islamic thinkers.[43] Allamah Muhammad Iqbal, a renowned twentieth-century Muslim poet from the Indian subcontinent, summarized Ibn Miskawayah's views on the origin of life. Iqbal explains that this Muslim scientist found that "plant life, in its lowest forms, does not "need any seed for birth or growth." The difference between this kind of plant life and minerals is "only in some little power of movement." This power of movement "grows in higher forms, and reveals itself further in that the plant spreads out its branches, and perpetuates its species by means of its seeds." Ibn Miskawayah believed that the last stage of plant development is the vine and the date palm, "where a clear sex differentiation appears." He maintains that the beginning stage of animality starts when plants are released from earth-rootedness, start to develop "the germ of conscious movement," and begin to acquire movement through the development of the senses of touch and sight, followed by the other senses. He believed that "[a]nimality reaches its perfection in the horse among the quadrupeds and the falcon in birds, and finally arrives at the frontier of humanity in the ape, which is just a degree below man in the scale of evolution." The final stages of the evolution of animals bring psychological changes with the growing power of discrimination and spirituality until humanity passes from barbarism to civilization.[44] Ibn Miskawayah believed and taught that plants, animals, and human beings belong to the same family which originated from simple life forms that gradually spread all over Mother Earth.

Abu Hassan al-Masudi (d. 957), a Muslim from Baghdad, was "the Pliny as well as Herodotus of the Muslim world."[45] He traveled extensively and gathered his gleanings into a thirty-volume encyclopedia in which he "surveyed omnivorously the geography, biology, history, customs, religion, science, philosophy, and literature of all lands from China to France." Al-Masudi summed up his scientific ideas in A Book of Information in which "he suggested an evolution from mineral to plant, from plant to animal, and animal to man."[46] Similarly, Uthman Amr al-Jahiz (776–868), a theologian and litterateur of Ethiopian origin, asserted that evolution occurred gradually, progressing from plant to animal to man.[47] Al-Jahiz's major works are Book of Animals (Kitab al-Hayawan) and Elegance of Expression and Clarity of Expression (Kitab al-Bayan we al-Tabyin). Muhammed al-Haitham (965–1039), a Muslim scientist from Egypt, was known to medieval Europe as Alhazen. Durant acquaints us with al-Haitham, "without whom we might have never heard of Bacon."[48] He is famous for his *Book of Optics (Kitab al-Manazir)*. He "studied the refraction of light through transparent media like air and water, and came so close to discovering the magnifying lens that Roger Bacon, Witelo, and other Europeans three centuries later based upon his work their own advances towards the microscope and the telescope."[49] Al-Haitham defended the doctrine of evolution. We do not have his original work, but we have two secondary sources. In one of them Ameer Ali, a historian, tells us that Muslim philosophers including al-Haitham accepted the theory of evolution. He summarizes their views on the subjects: "In the region of existing matter, the mineral kingdom, then animal kingdom, and finally the human being. By his body he belongs to the material world, but by his soul he appertains to spiritual or immaterial."[50] William Draper confirms al-Haitham's belief in evolution and his strong belief that "man, in his progress, [passed] through a definite succession of states" before he became the present day man.[51]

Ibrahim al-Nazzam, who lived about a thousand years before Darwin, was described as a "brilliant Muslim theologian of classical Islam."[52] He said that creation took place only once with infinite possibilities and potentialities. He taught that all elements that existed in the past or will exist in the future "were latent or potentially present in the stuff originally created—minerals, plants, animals, and men" and that they are "only the gradual realization of those latent potentialities."[53] Although early Muslim philosophers and scientists believed that the environment modified organisms and human beings, they held that changes in life forms were God's method to create and destroy. Hence, Ibn Khaldun concluded the paragraph on the origin of the human races (refer to chapter 3) with a quote from the Quran: "This is how God proceeds with His servants—And verily, you will not be able to change God's ways."[54]

Western scholars and scientists insist that Darwin formulated the theory of natural selection. Gould points out that within a decade, Darwin convinced the scientific world that evolution had occurred, but his theory of natural selection never achieved much popularity during his lifetime. It did not prevail until the 1940s.[55]

On the contrary, Muslims have long described natural selection working within the process of creation. Darwin borrowed the idea of artificial selection from horticulturists and farmers to explain natural selection. He pointed out that "when a race of plants is . . .well established, the seed-raisers do not pick out the best plants, but merely go over their seed-beds, and pull up the 'rogues,' as they call the plants that deviate from the proper standard." Darwin asserts that with the help of variations provided by Nature, humans apply the principles of selection to adapt organic beings to their use, but he also asserts that "natural Selection...is a power incessantly ready for action, and is immeasurably superior to man's feeble efforts."[56]

According to historical records almost ten centuries before Darwin, al-Biruni explained natural selection. He believed that God delegated active nature (kiyan) to perform its assigned duties in the universe. He explained natural selection in almost the same words as did Darwin centuries later. He also saw examples of selection in the methods of horticulturists, as well as in the natural behavior of bees. Al-Biruni wrote in Fi Tahqiq Ma Li'l-Hind (Al-Biruni's India):

> "The agriculturist selects his corn, letting grow as much as he requires, and tearing out the remainder. The forester leaves those branches, which he perceives to be excellent, whilst he cuts away all others. The bees kill those of their kind who only eat, but do not work in their beehive. Nature proceeds in a similar way."[57]

The following quotes from al-Biruni support the idea that long before Malthus and Darwin, he knew about the disparity between reproduction and survival. He describes here speciation and natural selection as result of structural advantages in some species of plants and animals. Al-Biruni wrote:

> "The life of the world depends upon sowing and procreating. Both processes increase in the course of time, and this increase is unlimited, while the world is limited. *When a class of plants or animals does not increase any more in its structure, and its peculiar kind is established as species of its own* [italics mine], when each individual of it does not simply come into existence once and

perish, but procreates a being like itself or several together, and not only once but several times, then this will, as single species of plants or animals, occupy the earth and spread itself and its kind over as much territory as it can find."[58]

Notice the similarity between the above quote from al-Biruni and the following passage from Darwin:

"A struggle for existence inevitably follows from the high rate at which all organic beings tend to increase [and] every organic being is striving to increase in a geometrical ratio; that each at some period of its life, during some season of the year, during each generation or at intervals, has to struggle for life and suffer great destruction. When we reflect on this struggle, we may console ourselves with full belief, that the war of nature is incessant...the vigorous, the healthy, and the happy survive and multiply [and] that any variation in the least degree injurious would be rigidly destroyed. This preservation of favourable individual differences and variations, and the destruction of those, which are injurious, I have called Natural Selection or the Survival of the Fittest."[59]

There is little difference in the views of al-Biruni and Darwin. Both agree that organic beings have unlimited ability to procreate, but that the world is limited. And both concur that individual variations in the structure of organic beings determine the survival of the peculiar kind among the species. Although Darwin labeled the natural phenomenon as "natural selection," al-Biruni closely described it when he used phrases such as "nature proceeds in a similar way," "the agriculturist selects" or "the forester preserves...while he cuts away all others." Materialists argue that the imperfections in nature suggest the absence of God in the evolution of life. Al-Biruni found that nature allows for the existence of evolutionary imperfections, citing the example of animals with imperfect limbs. He wrote:

"[If] Nature, whose task it is to preserve the species as they are, finds some superfluous substance...she forms [it] into some shape instead of throwing it away; when Nature does not find the substance by which to complete the form of that animal in conformity with the structure of the species to which it belongs ... she forms the animal in such a shape, so that the defect is made to lose its obnoxious character, and she gives it vital power as much as possible."[60]

Furthermore, like other Muslim scholars of his time, al-Biruni believed that while these imperfections may appear random to human beings, they are carried out by nature as a part of the "official function" designed for it by Allah.

"Frequently ... you find in the functions (actions) of Nature which it is her office to fulfill, some fault (some irregularity), but this only serves to show that the Creator who had designed something deviating from the general tenor of things is indefinitely sublime, beyond everything which we poor sinners may conceive and predicate Him."[61]

Regarding natural selection, Jalaluddin Rumi details the natural struggle for survival as the essence of life and a testimony of God's Will. He points out that the natural elements, such as air, earth, water, and fire, exist to serve God and that life is a universal struggle.

"The Universe, when you look at it, closely presents a universal struggle—atom struggling with atom like faith against infidelity. The struggle in action is the objective form of the principle of opposition, which has its basis in their inner nature. There is war in words and war in deeds and war in nature; between the parts of the Universe there is a terrible war. This war is the very constitution of the Universe; look into the elements and you understand it. Creation is based on opposition; therefore, every creature became warlike to get some benefit and avoid some injury. This struggle is not the phenomenon only of outward nature; even thy own self is a battlefield of one mental state opposing the other. The essence of soul transcends these oppositions; its nature is not (contradictory) like this; it is divine. Only opposites destroy opposites. Where there is no opposite there is eternal life."[62]

"Because without need Almighty God does not give anything to anyone; if there were no necessity, the seven heavens would not have stepped out of non-existence; the sun and the moon and the stars could not have come into existence without necessity; so necessity is the cause of all existence, and according to his necessity man is endowed with organs. Therefore, O needy one, Increase your need so that God's beneficence may be moved (to bestow new instruments of life on me)".[63]

Unlike modern hardcore evolutionists, such as Stephen Jay Gould and others, who believe in the absolute mastery of nature, al-Biruni and other Muslims see nature not as an absolute free agent, but as a force that operates in compliance with divine providence and whose actions are sometimes difficult to understand due to the limitations of the human mind. However, without God the evolver, man would not have come into existence. Gould accepts that if we were to rewind the videotape and replay the movie, the appearance of humankind on the earth is unlikely.[64] This is probably one of the reasons that Muslim philosophers and scientists believed that without divine guidance for the imperfect nature, man would not have evolved. Muslims accept that all created beings are imperfect in structure and wisdom. The only one who is perfect is God, and so He is called Al-Hakim (the Perfect in Wisdom).

Based on the pre-Darwin Muslim works that have been cited, we can conclude that Muslim scholars viewed life as a bush and all life forms as either twigs or branches of that bush. They believed that humans are related to apes as apes are related to their predecessors. Al-Biruni's and Rumi's explanations of natural selection were rendered almost verbatim by Darwin centuries later. Yet Western historians scarcely mention Muslim names when they record the history of evolution.

The Western claim that Aristotle is the originator of the ladder of nature is an inference that cannot be corroborated because "not a single original word that Aristotle wrote exists today."[65] By contrast, we have detailed descriptions of the ladder of nature in Muslim works. It is unlikely that Muslims derived their evolutionary theories from Judeo-Christian teachings or from Darwin and Spencer. The abundant evidence presented here demonstrates that Muslims are the originators of the theory of evolution, and William Draper is correct when he calls it the Muslim Theory of Evolution. The only difference one can cite between the Muslim scholars and Darwin is that Muslims believed that the existence of the ladder of nature was the result of divine will and providence.

Why did Muslim scholars develop the evolutionary paradigm of the origin of life, rather than accept the existing belief of their time? They derived their thought from their divine book, the Quran, in addition to their independent scientific observations. These inquisitive but believing Muslims took seriously the following commands in the Quran as a methodology for a correct understanding of the world:

> **51:20–21** On the earth are the signs for those who have sure faith (in the meaningfulness of all things), as also (there are signs) in your own self: will you not, then, observe?

The Quran further commands us to search and learn from the earth the process of the creation of life, man, and the universe (29:20). Early Muslims considered the universe as another book of God, and in trying to decipher the laws of nature; they were also following God's commands.

The above verses suggest that the universe, with all its physical, chemical, biological, and psychological laws, can be comprehended. They also suggest that every Muslim should understand God's method of creation. What is the situation with Muslims of today? It is a sad story! Instead of intellectual pursuit enjoined by God's guidance, intellectual oppression is the rule in most Muslim countries. The Prophet said, "One learned man is harder of the devil than a thousand ignorant worshippers," and "The ink of the scholar is holier than the blood of a martyr."[66] Unfortunately, the sacred ink has dried up and unholy blood is dripping from Muslim pens in many part of the world, including at the site of the World Trade Center. I believe that there exists an urgent need for Muslims to reacquaint themselves with the meanings of the Quran. Therefore, let us next review the Quran to find out what it says about the origin of man.

CHAPTER 7

The Quran
and the Origin of Man

THE QURAN does not have a chapter on the genesis of human life, as do the Jewish and Christian scriptures. Christians and Jews believe that God created human beings in His image and created Eve from Adam's rib. Most contemporary Muslims also believe that God created Eve from Adam's rib, although not a single verse in the Quran supports this belief. What, then, does the

Quran say about the creation of human beings? The Quran is a book of guidance that asks its believers to investigate and understand nature. It does not spoon-feed them with knowledge, but advises them to observe and reflect on nature. An example of this method of teaching can be observed in the following verses:

> **3:190–191** Truly in the creation of heavens and the earth, and in the alternations of the night and day are signs for those who have acumen, who utter (the name of) God, standing, sitting, and on their sides, and ponder over the creation of the heavens and the

earth, (saying), "Our Lord, You have not created these in futility. Glory be to You: guard us then from the torment of the fire.

51:20–21 On the earth are signs for those who have sure faith (in the meaningfulness of all things), as also (there are signs) in your own self: will ye not, then, observe?

The verses instruct that contemplating God's creation, pondering the open book of the universe, observing God's creative hand moving this universe, turning the pages of a book, and learning from human anatomy all evoke true worship and remembrance of God. This verse and other verses encourage Muslims to examine the world with a scientist's eyes to understand the products and process of creation. Thus, our perception of the process of creation is enriched as we continue to study all pertinent verses of the Quran and to explore the universe as new technological means become available to us.

Verses about the origin of life and man are scattered throughout the Quran; we have collected the key verses in this chapter in order to develop a clear picture of the meanings they convey. A casual reading of the verses may give us the impression that they contradict each other, but apparent contradictions evaporate when we interpret the meanings of all the verses collectively rather than the meaning of each single verse. More importantly, our conclusions should be consistent with our overall reading of the Quran. For Muslims, the internal consistency of the Quran is the proof of its divine origin:

4:82 Do they not consider the Quran (with care)? Had it been from other than God, they would have found therein much discrepancy.

Thus, Muslims generally contend with verses that seem contradictory in one of three ways: They conclude either that their understanding is wrong, or that the totality of knowledge at a given time and place is not sufficient for them to understand the purpose and message of the verses, or that the verse is allegorical.

The Quran enlightens us that all things are created by God in accordance with His grand design, that God is manifested through His

creations, and that His attributes are the link between human beings and their comprehension of the Divine.[1] Out of the titles and
ninety-nine attributes ascribed to God in the Quran, four grant us
insight into His process of creation. Those key attributes are *Rabb*
(the Sustainer), *al-Khaliq* (the Creator), al-*Baari* (the Evolver), and
al-Musawwir (the Bestower of Forms).

The first command that came to the Prophet was:

> **96:1–2** Read in the name of thy Sustainer (Rabb), who has cre-
> ated created man out of a germ-cell.

The general translation of the word *Rabb* as Sustainer does not
convey its overall meaning. The noun *Rabb* is derived from the
Arabic word *Rububiyat*, the meaning of which cannot be fully rendered in English with one word. Based on his analysis of the works
of early Arab lexicographers, Abul Kalam Azad, a Muslim Quranic
scholar and commentator from the Indian subcontinent, deciphers
the meaning of the word as follows:

> "To develop a thing, stage by stage, in accordance with its inher-
> ent aptitude and needs, its different aspects of existence and also
> in the manner affording the requisite freedom for it to attain its
> full stature."[2]

Likewise, Imam Aul'l-Qasim ar-Raghib (11th century), in his
book *Al-Mufradat fi Gharib al-Quran* (*The Vocabulary of The Quran*),
defines the meaning of the word *Rabb* as follows:

> "*Rabb* signifies the fostering of a thing in such a manner as to
> make it attain one condition after another until it reaches its goal
> of perfection."[3]

The major components of the meaning of the word *Rububiyat* are
thus: (a) development of a thing by an external agent, (b) a step-by-
step process, not an instant event, and (c) the freedom for the objects
"to attain full stature" within the overall creative process. Therefore,
Rabb, the derived noun from *Rububiyat*, means an evolver. The use

of the noun *Rabb* as an attribute of God suggests that God lets organisms evolve, affording them the freedom to attain complete perfection within the limits of His laws of nature.

Another descriptive term of God in the Quran is *al-Khaliq*. The word is derived from the Arabic root verb *khalaqa*, which is used in nearly all the verses pertaining to the creation of human beings and the universe as a whole. To explain how the Quran describes this process, we first need to understand the meaning of *khalaqa*. Almost all commentators and translators of the Quran, both Muslim and non-Muslim, translate the meaning of this verb as corresponding to the English verb "to create." However, this translation does not convey the original and complete meaning of the verb. The Arabic-English lexicon compiled in 1883 by a British lexicographer, Edward William Lane, guides non-Arabic speaking students of the Quran to understand its meanings. At the request of the Duke of Northumberland, Lane edited his lexicon after twenty-five years of exhaustive study of the Arabic language. He explains that the original Arabic language of Ma'ad, during the time of Muhammad, was a highly complex and difficult linguistic system. As a result of Muslim conquests and the intermingling of Arab culture with that of other nations, the classical Arabic language lost its original character and assumed a simpler version of the original language that became predominant even in Arabia.[4] Lane's lexicon is the most authoritative for the non-Arabic speaking students of Islam because he traced the original meanings of the words back to the time of Ma'ad and the Prophet Muhammad. His interpretation of Arabic words is based on earlier lexicographers and grammarians, such as Al-Khaleel (author of Eyn, died in 667 a.d.), Esh-Sheybanee (author of Jeem, Nawadir, and El-Ghareeb el–Musannaf, died in 802 a.d.), Al-Baydawee (author of Exposition of the Quran, 1290 a.d.), Al-Feiyoomee (1333 a.d.), al-Firuzabeedi (author of Kamoos, 1330–1421 a.d.), and Seyyid Murtada-al-Zabeedi (author of Taj-al-Aroos, 1732–1791 a.d.).[5]

According to Lane, the verb *khalaqa* means "proportioning a thing into another thing" and "to bring a thing into existence according to a certain measure, or proportion, and so as to make it equal to (another thing)."[6] In "proportioning a thing into" an original and previously non-existing thing, it also signifies "the originating, or to bring a thing into existence after it had not been, or the bringing a thing into existence from a state of nonexistence."[7] Hence, there are three components to the meaning of the word *khalaqa*: (a) shaping an original substance or entity into another object ("proportioning a thing to another thing"), (b) the newly formed object or creature must have its own peculiar characteristics so that it can be identified separately from its original source—not like father-to-son, but like a tree to a canoe or clay to a human, and (c) the newly formed object or creature with its characteristic features were nonexistent before its original birth ("to bring a thing into existence from a state of non-existence"). In this process the new creature becomes a prototype. Therefore, the classical meaning of the Arabic word khalaqa can be summarized as follows:

> "To bring a thing into existence according to a certain measure, or proportion, so as to make it equal to another thing that is not pre-existing."[8]

The above meaning of *khalaqa* is more befitting with the overall reading of the Quran because it is based on the true language that was prevalent among the natives of the Arabian Peninsula at the time of the Prophet Muhammad and not upon modern Arabic. Moreover, it does not contradict what we observe in nature, because the meaning of the verb *khalaqa* implies that

various life forms were not created simultaneously on the earth, but in stages. Now that we understand the true meaning of *Rabb* and *khalaqa*, in order to gain a better understanding of the way the Quran describes the process of creation, those of us who read English translations of the Quran should substitute the above meanings wherever the verbs *create* or fashion appear.

The Quran describes God as "the Evolver" (*al-Baari*) and "the Bestower of forms" (*al-Musawwir*), reinforcing the idea of creation as a stage-by-stage process. The verse reads:

> **59:24** He is God, the Creator [al-Khaliq], the Evolver [al-Baari] and the Bestower of forms [al-Musawwir]

Al-Baari, the Arabic word in the above verse, is derived from the verb *baara,* which means "a thing's becoming clear, or free, of, or from another thing; either by being released [therefrom]" or by "evolution from a previously created matter or state."[9] So God, the executor of such evolution, is *al-Baari* (the Evolver).[10] The word *al-Musawwir* is derived from the verb *sawwara,* which means to sculpt a thing and give definite form or color to make things exactly suitable for a certain end or object.[11] Hence God is called *al-Musawwir,* or the "sculptor, of all existing things, who has established them, given to every one of them a special form and a particular manner of being whereby it is distinguished, with their variety and multitude."[12]

We would not be wrong to infer, then, that if God is the evolver, He is also the creator of new life forms from those which He had previously created. Moreover, since He is the "Bestower of forms and colors" so that His creations function suitably in environments to achieve "certain ends or objects," we could logically infer that by changing the genetic configuration, He creates new species capable of adapting to the changes in environment over the various ages of the earth. The Quran emphasizes that such changes are the result of the will of God. In summary, the meanings of *Rabb, al-Khaliq, al-Baari,* and *al-Musawwir* confirm the belief that evolution and creation are not contradictory but mutually complimentary. So, based on the attributes of God, creation is a divine process and the theory of evolution is the human understanding of the process. In biology, "genotype" denotes the total genetic code of a living creature and "phenotype" denotes the outward physical appearance of a living creature as a manifestation of genotype. By applying these modern biological terms in the context of our discussion, if creation is genotype, the theory of evolution is phenotype.

We learn from the Quran that the creation of the universe and its contents was not an instant event, but a process: "It is He who beginneth the process of creation (*khalaq*) and repeats it." (10:4) Yusuf Ali, a well-known Muslim translator, explains, "God's creation is not a simple act, once done and finished with. It is continuous, and there are many stages, not the least important of which is the Hereafter, when the fruits of our life will be achieved."[13]

In support of this inference we cite:

32:7 He who has made everything which He has created most good: He began the creation (khalaq) of man with (nothing more than) clay, and made his progeny from a quintessence of a fluid despised.

76:2 Verily We created man from a sperm yoked (to the ovum).

The creation of the Prototype human species began from clay. However, their progenies were created by an "emitted drop of semen" "yoked with ovum." Through scientific evidence, we know that the creation or birth of a human baby involves the evolution of cells through a process of meiosis, during which the nucleus of a cell divides and forms a mature germ cell comprised of sperm and ovum that unite in a continuous process of reproduction. Although the Quran does not render an exact explanation of the above, its verses lead us to infer that the evolution of yoked sperm and ova through all later stages of intrauterine fetal growth to form human progeny does not occur instantaneously, but as a step-by-step process rooted in finite time and space in relation to an earthbound person's time frame. Similarly, the creation of prototype Homo sapiens began with clay. The phrase in the verse 32:7, "He *began* the creation of man (emphasis added)" indicates that the beginning was not an end in itself. Additional steps necessarily followed to complete the intended task.

Other verses also clearly imply a time lapse. For example: "He fashioned *(khalaqa)* you from sounding clay, like unto pottery."[14] (55:14). The word *fakha'ar,* in Arabic text of the Quran, means "baked pottery or a baked vessel of clay."[15] We can read this verse as follows: "Just as clay is molded into shape in stages and baked into pottery, prototype human species was also created through successive stages over a period of earthly time."

This inference is supported in the following verses: "He fashioned you, and perfected your shapes" (64:3) and "He created *(khalaqa)* you in successive stages" (71:14). The latter verse unequivocally states that the creation of mankind was not a magical *ex nihilo* instant event but a step-by-step transformation. Today, we scientifically explain these stages as a process that expands from nothingness to the Big Bang, from the Big Bang to the earth, from clay to nucleotide, to deoxyribonucleic acid (DNA) to microorganism, from microorganism to marine animals, from marine animal to Dryopithecus, from Dryopithecus to Australopithecus to Homo erectus to Homo ergaster. Homo ergaster branched into Neanderthal and Cro-Magnon man. Cro-Magnon evolved into modern man. Modern agnostic biologists describe the "creation by stages" as follows:

> "Evolution usually proceeds by "speciation"—the splitting of one lineage from a parental stock—not by slow and steady transformation of these large parental stocks. Repeated episodes of speciation produce abush."[16]

According to the Quran, man is created from components:

> **82:7–8** Who created (*khalaqa*) you, then proportioned you into whatever form He willed. He made you out of components [*rakkaba*].

The verb *rakkaba* in the original Arabic text of this verse means "to create a thing from components; put or set one part of it upon another."[17] Therefore, the verse indicates that the necessary components were created as prerequisites for the creation of man.

Even though the Quran states that God created humans in successive stages, it does not describe all the stages of their evolution from clay before they became Homo sapien sapiens (modern man). Sperm and ova are not human. The forty-six chromosomes present in the fertilized human egg are the encapsulated secret codes of a future human being, but we do use the terms "man" or "woman" to describe them. They are still a bundle of nucleotides, phosphate, and sugar. A thing is named only when it comes into its own form and acquires specific characteristics. The stages through which human beings came into existence have their own names—water, clay, soil, microorganism, marine animals, mammal-like reptiles, Dryopithecus, Australopithecus, Homo erectus, Homo ergaster, and Cro-Magnon man. However, all those are only landmarks in the journey of the creation of mankind. These landmarks occurred a long time ago, and memories of those stages do not remain in the human brain but are stored in the vestigial structures, fossils, biochemistry, immunology, and so on. Those unremembered stages of evolution in the creation of mankind might be what the following verse alludes to:

> **76:1** Has there come on man a while of time when he was a thing unremembered?

The verse suggests that mankind existed in some kind of form that was not recognizable as human. Perhaps this was one among many verses that may have inspired Jalaluddin Rumi (1207–1273), a great Muslim sage, to compose the strophe quoted in the previous chapter, in which he describes the emergence of human species from the crucible of evolution without remembering his earlier forms of existence.[18]

The Quran reads: He said, "So it shall be." Your Lord said, "Easy it is for Me, as I created you before you were nothing." (19:9). This verse explains that mankind and the universe were created from nothingness. The initial void, with the exception of the existence of God, was nothingness—no matter, space or time. When the Quran refers to a state of nothingness, it is not saying that human beings were created *ex nihilo* without any connection to other life forms or species. If Muslims do not accept this inference, then they have to admit that the Quranic verse 19: 9 is contradictory to many other verses in the Quran. They need to ask the question: If man was created *ex nihilo*, why does the Quran state that man was created from water, clay, soil, and germ-cell, and in successive stages?

Almost all biologists accept the fact that life originated in water. This is what the Quran reveals about the aquatic origin of life:

> **21:30** Do not the unbelievers see that heaven and earth were joined together (as one unit of creation), before We clove them asunder? We made from water every living thing. Will they not then believe?

The verse points out that the universe was in a condensed state. In modern scientific terminology, that state is called a singularity.

The term refers to the moment when time and space merged into one entity. Astronomers use the term Big Bang to describe that initial defining moment of creation of the universe. Following that, God created life from water. "Every living thing (*minal-maa-'i kulla shay-'in hayy)*" in the above verse certainly includes mankind. Therefore, the early stage of the evolution of man occurred in water.

A sampling of some Quranic verses confirms the idea that humans are linked to both organic and inorganic realms:

> 23:12 We fashioned man from quintessence" *(sulalah)* of clay.

The word *sulalah*, in the Arabic text of verse 23:12, means "an extract of a thing; the clear, or pure, part or choice, or best, or most excellent part of a thing."[19] One may conclude that man was created from an extract of clay. Does *sulalah* mean silicone? We have learned previously that at least some scientists induct clay minerals as a catalyst, as well as a stabilizer, in the polymerization reaction of amino acids to RNA or DNA. Muslims believe that Quranic verses cannot contradict each other on any particular issue. The Quran states in various verses that man was created from water, clay, quintessence of clay, and so on, and we know that these substances are different in chemical and physical constitution. Unless we consider these substances as part of different stages in the early creation of humans, then the Quranic verses do contradict each other. If the prevalent contemporary Muslim belief in an instant creation without intermediate stages is taken as the truth, the contemporary *Ullamah* (Muslim religious scholars) must propose credible explanations for these conflicting verses.

Quranic verse 24:45 suggests that mankind is a part of the animal kingdom:

> **24:45** "And God has created *(khalaqa)* every living animal from water: Of them some that creep on their bellies: Some walk on two legs: Some walk on four: God created *(khalaqa)* what He wills: For verily God has power over all things…"

(*Wa-Allaahu khalaqa kulla daaabbatin min maa': fa-minhum-mân yamshy 'alaa batnih; wa minhum mân yamshy 'alaa rijlayn; wa minhum mân yamshy 'alaa'arba'...*)

Verse 21:30 uses the phrase "every living thing (*minal maa' kulla shay-'in hayy*)." Here, verse 24:45 offers additional clarification: "every animal (*kulla daaabatin min maa'*)." This verse specifies that animals were created out of water and explains the categories included among animals. Humans and sometimes apes are the only living creatures that always walk on two legs, while birds can fly and walk on two legs. Therefore the two legged animals mentioned in verse 24:45 can refer to birds or apes or human beings.

The grammatical structure of the above verse with the phrase *fa min hum* is highly significant. If the noun *dabbah* (animals) were applied only to rational or irrational creatures separately, the two phrases, *fa min-hunna* or *fa-min-ha*, would have been used in proper Arabic grammar. Instead, the Quranic use of the phrase *famin hum* in the verse conveys the Arabic noun *dabbah* (animals) in the verse refer to both rational and irrational creatures.[20] Therefore, the verse states that a rational animal that walks on two legs was also created from water. Humans are the only creature that walks on two legs all the time. Moreover, humans are the most rational of all creatures. Therefore, the rational animal that walks on two legs referred to in the verse is a human, so humans belong to the animal kingdom. Moreover, five hundred years before Darwin, the Muslim scholar, Ibn Khaldun wrote:

> "[M]an belongs to the genus of animals" and that "God distinguished from them by ability to think, which He gave man and through which man is able to arrange his actions in an orderly manner."[21]

The following verses clarify further that humans were created through transformation/evolution.

82:6–7 O you man, what deluded you concerning your Munificent Lord, who created you (خلق *khalaqa*) fashioned you (سَوَّى *sawa*) and shaped you in perfection (عدل *adala*).

If God created humans with no connection to pre-human species including other hominids, the Quran would not have said, "We created you (*khalaqa*), then fashioned you (*sawa*)." Similarly, if humans did not evolve in stages but served as a perfect creation, then the verb *sawa* in the verses becomes superfluous because there is nothing to perfect in a perfect being.

The commentator translates the verb *adala* (عدل) as "shaped you in perfection." Lane's lexicon gives another meaning to this word: "rate a thing as equal to a thing of another kind so as to make it like the latter." Therefore, it means to transform a thing into another, so that latter is distinctly identifiable from the former. In support of this meaning, he quotes the following Quranic verse: "All praise be to God who created the heavens and the earth, and ordained darkness and light. Yet the unbelievers make others equal {*adala* (عدل)} of their Lord (6:1)." Lane gives the phrase فلان بفلان عدل (*fulaanun bifulaan-in 'adala*), which means, "He made such a one to be equal, or like, such a one."[22] Therefore, the verses 82:6–7 must be rephrased as: "O you man, what deluded you concerning your Munificent Lord, who shaped you from a pre-existing thing (*khalaqa*), sculpted you (*sawa*), and transformed you to [distinctly identifiable] perfect human shape [i.e., Homo sapien sapiens] (عدل *adala*)."

The Quran addresses mankind, proclaiming that God created both mankind and also "those who came before" mankind: "Oh! Mankind! Worship your Lord, Who hath created you and those before you, so that ye may ward off (evil)." (2:21). Hence, it tells us that there existed other people before the emergence of modern man (Homo sapiens). We call them Homo erectus, Homo ergaster, Neanderthal man, Cro-Magnon man, etc. but six hundred years before Darwin, the great Muslim Sufi Ibn Arabi called them "the animal men." Most Muslims interpret the above verse in a more limited fashion, believing that the term "mankind" means people living at the time of the Prophet, and that "the people before you" refers to the ancestors of the contemporaries of the Prophet.

However the best explanation of a verse in the Quran is by another verse. We have learned earlier the divine epithet al-Musawwir (sculptor) could also mean, in current scientific vernacular, someone who creates a distinct species. Therefore, al-Musawwir created perfect human beings sculpted from a previously created hominid species. Verse 6:133 gives explicit evidence that modern humans were evolved from earlier hominids and explains the Quranic verse 2:21 the way as it is explained above. "Thy Lord is All-sufficient, Merciful. If He will, He can put you away, and leave after you, to succeed you, what He will, as He produced you (أنشأ) from the seed of another people." (6:133) The context in verse 130 of the same chapter shows that it is addressed to mankind. Almost all translators of the Quran render the Arabic word `ansha (أنشأ) as "raised." This is not a correct translation of the word, however. Lane translates it as follows: "originated it; brought it into being or existence; made it, or produced it for the first time, it not having been before."[23]

Initially, verse 6:133 states that God may replace modern humans with another species ("whomever He pleases to succeed you"). The process of the creation of the new species (khalifah) from human seed is similar to when God "produced humanity into existence for the first time (أنشأ)." The verse also suggests that the human species was originally created from "the seeds of another people," i.e. that modern humans were evolved from earlier people. Therefore, verse 6:133 can be rephrased as: "Thy Lord is all Self-sufficient, Merciful. If He will, He can put you away, and leave after you, to succeed you, what He will, as He originally created you (أنشأ) [as a species] from the seed of another people."

A respected twelfth-century exegete of the Quran, Fakhr al-Din al-Razi (1149–1209), explains the verse similarly: "[God] said: 'As We created you from the seeds of a different people.' For a wise person, were he to contemplate on this statement...he would know that the Almighty created mankind from a sperm; a sperm that did not contain his picture in any form or way."[24] If we replace the word "picture" with our modern scientific terminology "genotype" in the above quote from Razi, his explanation of evolution of human from earlier species become more clear and evident: "[God] said: 'As We created you from the seeds of a different people.' For a wise person, were he to contemplate on this statement... he would know that the Almighty created mankind from a sperm; a sperm that did not contain his [genotype] in any form or way."

I translated the word, *khalifah*, in the above verse 6:133 as "species." However, most Muslim commentators of the Quran translate the noun *khalifah*, in the above verse, as vicegerent. Human as vicegerent means that the human species as a whole is the deputy appointed to act on the authority of a God, especially in administrative duties. This word, *khalifah*, is actually derived from the verb *khalafah*, which means "one entity that succeeds another or remains after another that has perished or died."[25] Therefore, the literal translation of the noun *khalifah* is "someone remains after another that has perished or died," but the term vicegerent is also correct, because human supremacy over the earth was established by divine revelation of all knowledge to human beings through the Prophet Adam. According to biology, *Homo sapiens* as a species "succeeded or remained after" the earlier hominid inhabitants of the earth "that had perished or died." Therefore, logically a new species is a *khalifah*. We have paleontological data that pre-modern human species existed and perished with the emergence of modern humankind. This early hominid was identified by Ibn-Arabi (1165–1240) and other early Muslims as an animal-like man, but modern biologists called him *Homo erectus*, Cro-Magnon, or other such names.

Furthermore, according to the following verses, God may create new species transfiguring Homo sapiens: "You shall surely travel from stage to stage." (84:19). "We may transfigure you and make you what you know not." (56:61). Additionally, the following verse predicts a newly created species able to verbally communicate with human beings: "And when the word is fulfilled concerning them (mankind), We shall bring forth a creature of earth to speak unto them because mankind has no faith in Our Revelation." (27:82). Clearly, the following verse indicates that the survival or extinction of any life forms is subject to God's will: "Thy Lord does create and choose as He pleases." (28:68)

Geology and paleontology occupy a key position in evolutionary studies. We have quoted earlier, in 29:20, that the Quran counsels Muslims to study the earth and its contents to learn the divine process of creation. The verse encourages mankind to learn to create. We have earlier offered historical documentation showing that Muslim scholars undertook the study of geology and fossils long before Europeans started to do so. We have seen that al-Biruni observed that some hominid species were much bigger in size than modern man. He based his conclusion on his observation of the size and construction of ancient caves and the bones buried in them.

One can find many points of harmony between modern science, the Quran, and these early Muslims: First, the Quran and at least some scientists agree that clay played a role as a catalyst in the origin of life and man, and that life began in water. Second, they agree that life is a bush with many branches and man is only a small twig on that bush. Third, they both confirm that the origin of life and its development has been a gradual process, not an instantaneous event in our time frame. Both geology and paleontology provide information on the origin of life and its progressive changes, and they declare that human beings belong to the animal kingdom, that the original life form was gradually transformed into new forms, and that this transformation occurs with the participation of "the office of active nature (kiyan)." They also suggest that, in the future, human beings may be transfigured into a different form of life, the shape of which is not known to us at this time. The following verses connect all the above verses together to show how God bound all life forms together into a whole:

> **87:1–3** Extol the limitless glory of thy *Rabb:* [the glory of] the All-Highest, who creates (everything), and thereupon forms it in accordance with what it is meant to be, and who determines the nature (of all that exists), and thereupon guides it (towards its fulfillment).

In the above verses, the Quran commands the Muslims to glorify their *Rabb*, who creates everything and then lets His creations advance to perfection under His guidance. The word guidance has to be understood within the context of Islamic metaphysics of the future. Within this metaphysics, guidance means that God provides possibilities or proposals to each stage or component of the universe from which to choose and by which to realize their next state. All changes at every level are gradual in our framework of time and space. The creatures' selective actualizations of the options suggested by God result in visible monuments of divine creation. (See CHAPTER 1 on the Islamic metaphysics of the future.) When genes accept a suggested possibility or possibilities from Allah, a new species is born or created. We will discuss this aspect in detail later.

Within the logic of this metaphysics, we can assume that variations of natural species that appear seemingly without design are in fact the results of divine will and grand design. This is not an attempt to fit the Quran to modern scientific reasoning; on the contrary, nine hundred years before Charles Darwin, al-Biruni understood the ways of God and how He brings about changes in life forms, as presented above. In Athar-ul-Bakiya (Chronology of Ancient Nations), he asserted that natural changes in life forms are decreed by Divine will and "show that the Creator, who had designed something deviating from the general tenor of things, is infinitely sublime, beyond everything which we poor sinners may conceive and predicate of Him."[26] Furthermore, al-Jahiz (776–869) wrote:

> "Individual defects in Nature do not lead us to the conclusion that all natural things happen by accident and chance. Thus the claim, founded on individual cases, that things happen by accident is incorrect and betrays ignorance. If you object, 'Why should such cases occur?' we reply that things are not determined by Nature so that the only explanation is, contrary to the claims of some people, that they are permitted by design of the Creator. He made the Nature run its normal course in most cases, deviating sometimes, however, because of certain accidents; this fact implies that Nature is commissioned and designed, needing the will of the Creator to reach its purpose and achieve its works."[27]

In modern scientific terms, the description of "something deviating from the general tenor of things" by al-Biruni or "Nature...deviating sometimes" by al-Jahiz can be read as the process of mutation. Later we shall discuss how the elements of chance and accidents merge with creation. All life forms on earth, including virus and vampire, man and monkey, grass and grasshopper, are the end result of the differences in the number and sequence of the four basic nucleotides (adenine cytosine, guanine, or thymine) made out of one nitrogenous base, one 5-carbon sugar (deoxyribose), and one phosphoric acid residue. The genetic structure of the chromosomes determines the minutest physical, chemical, and immunological characteristics of various species in both the animal and plant kingdoms. Adding the meanings of *Rabb, al-Khaliq, al-Baari,* and *al-Musawwir* to our knowledge of paleontology, homology, analogy, molecular biology, and other branches of science, we may conclude the following about the matter of creation: God, through His proposals, proportioned and shaped the number and sequence of the four nucleotides in the genes to create new forms of life from previously existing creatures. The Quran commands Muslims to learn the about process of creation by observing and decoding its visible signs on the earth and in their own selves (51:20–21 and 29:20). The early Muslims followed these commands in their exploration of nature. In classical times centuries before Darwin, Muslims believed that the creation of mankind began with clay and water and progressed through the lower life forms to Homo sapiens. Western scientists have only recently accumulated additional empirical data from the earth (fossils), from morphology, from the biochemistry of all organisms, and from other sources. This data supports Muslim thought regarding how God makes (*khalaqa*) new life forms from nonexistence. Therefore, observation of the evolution of life and the Quranic teachings of creation by God are not contradictory but complimentary.

Unfortunately, Western scientists who grew up in Judeo-Christian traditions and who did not agree with the literal meaning of Genesis were forced to interpret the scientific data differently. The study of comparative anatomy led to the concept that there are two types of similarities between living organisms. The first is analogous resemblance, where a fundamentally dissimilar structure has been modified or adapted to serve similar ends. For example, the resemblance between the surfaces of the wings of a fly, a bird, or a bat is analogical, but they are formed from different materials. The insect wing is simply a membrane connected by veins, while all vertebrate wings are created from the typical bones of the forelimb of a land vertebrate.

The second, homologous resemblance, occurs when a fundamentally similar structure is modified to serve dissimilar ends. The hind limbs of a horse, a bat, or a whale, have bones that are modified in size, in details of shape, by reduction, or even by fusion of bones (as in the horse) to serve the needs of the different species. (See Figure 5-3.)

Evolutionists point out that analogy and homology provide unequivocal evidence that evolution occurs without supernatural intervention. But if read in the light of the literal meanings of the Arabic verb *khalaqa* and God's epithets *al-Baari, Rabb,* and *al-Musawwir,* analogy and homology would point to a divine act of creation by proportioning one thing to another. Creation and evolution are integral parts of a divine act. Interpretations of analogy and homology vary depending upon the interpreters. Atheistic evolutionists consider them evolutionary processes detached from divine intervention, while Muslims, in the classical period, considered them willful acts of divine creation. In the light of Islamic metaphysics of the future, we can see analogy and homology as the results of some creatures' acceptance of divinely suggested possibilities arriving through the messenger moments of the future.

One of the cardinal principles of the modern synthetic theory of evolution is the slow accumulation of small mutations. In excerpts quoted in the previous chapter, Ibn Khaldun expresses the cardinal principle of Neo-Darwinism in terminology used in his time. He describes evolutionary processes as the "connection and full preparation to become the next stage," "the last stage of each group is fully prepared to become the first stage of the next group."[28] Such statements can be paraphrased using modern synonyms as "the accumulation of small mutations results in speciation."

When Judeo-Christian theologians started to present evolution as a divine plan slowly unfolding through the ages, Bertrand Russell, a pacifist philosopher, wondered whether, during such a process "the omnipotence (God) [was] quietly waiting for the ultimate emergence of man?"[29] Similarly, many traditional Muslims question the truthfulness of the stage-by-stage progression of creation because they observe the literal meaning of the Quranic verse: "Verily, when He intends a thing His command is 'Be', and it is." (36:82). The theory of relativity provides answers to these important concerns. Einstein convinced the world that time is not absolute but relative. He demonstrated that each observer carries around his or her own personal time, which does not generally agree with anybody else's.[30] Time is not absolute; it is elastic. (Refer to Chapter 2). Therefore, for God, our experience of events and characters in the drama of life is compressed into "a twinkling of an eyelid." So, after the first spark of life 3.5 billion years ago, as far as God is concerned the human appearance on earth is right ontime. This is depicted in Figure 7-1.

In the ninth century, al-Biruni argued that the age of the universe could not be calculated. He asserts, like our modern physicists, that for an observer on earth "the colossal and minute moments of creation seem to have formed over a long period of time."[31] The events in the process of the creation of life are current events in the timeframe of God, but the earthbound man experiences creation as a multitude of events and evolution occurring over a period of 3.5 billion years.

When God orders Muslims to "travel around the earth and see how [He] originated creation" (29:20), we do so knowing that the process of creation is comprehensible to us only through our earthly experiences. Had God said one thing in the Quran and given evidence of another, we would have cause for doubting His existence. But He has provided us with evidence, tools, and the ability to discover and understand His creation. By exploring and comprehending that process, we confirm our belief in His existence and are able to accept the assertion that human existence occurred as a result of His command.

Sadly, many contemporary Muslims do not realize when they view life as a movie that they are in it. They may accept God as the Creator, but they assume that their creation has no connection with any primates. They may not realize that their lives constitute one frame of the movie of life, and that there are other frames in front of and behind them. They claim that they are the whole movie. They begin and end with themselves. However, God's epithet, the Evolver (al-Baari), within space-time relativity suggests the opposite. (Figure 7-2.)

Similarly, materialists among scientists who can see most of the movie of life remain satisfied with the concept of natural selection to explain creation. We have seen earlier that Pre-Darwinian Muslims described an active nature (kiyan). They believed that the active

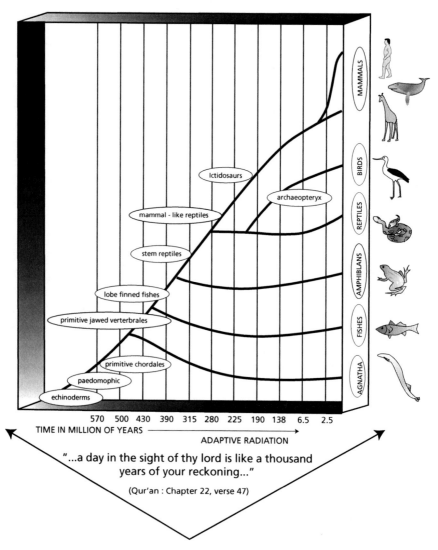

"...a day in the sight of thy lord is like a thousand years of your reckoning..."

(Qur'an : Chapter 22, verse 47)

A moment of Al-Baari (The Evolver)

7-1. EVOLUTION, CREATION, THE THEORY OF RELATIVITY, AND AL-BAARI (THE EVOLVER).

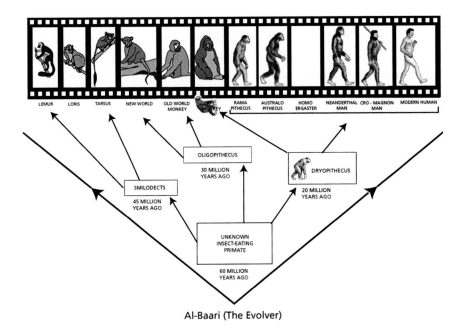

LEMUR · LORIS · TARSUS · NEW WORLD MONKEY · OLD WORLD MONKEY · RAMA PITHECUS · AUSTRALO PITHECUS · HOMO ERGASTER · NEANDERTHAL MAN · CRO - MAGNON MAN · MODERN HUMAN

OLIGOPITHECUS

30 MILLION YEARS AGO

DRYOPITHECUS

SMILODECTS

20 MILLION YEARS AGO

45 MILLION YEARS AGO

UNKNOWN INSECT-EATING PRIMATE

60 MILLION YEARS AGO

Al-Baari (The Evolver)

FIGURE 7-2. AL-BAARI (THE EVOLVER), SPACE-TIME RELATIVITY, AND THE EVOLUTION OF MAN FROM AN INSECT-EATING MAMMAL.

nature is endowed with a faculty to select species as the farmer selects the better seed. They believed also that its laws submit to the will of God. Materialists accept the existence of natural laws—but they evade the question of why natural laws exist.

When we look around us, we know where tangible things are—where the United States is, where Mecca is. We recognize the position of moving objects (cars, trains) at any particular moment. When we first extended our experience into the atomic world and drew a picture of an atom as a micro-version of our solar system, we assumed that the internal constituents of the atom must rotate just as precisely as the planets around the Sun. However, we now see that when we descend the stairway into the atomic world, we lose our sense of certainty about everything and find ourselves in a fuzzy and chaotic world.

An electron does not appear to follow a meaningful, well-defined trajectory at all. All known subatomic particles—even whole atoms—"cannot be pinned down to a specific motion. Scrutinized in detail, the concrete matter of daily experience dissolves in a maelstrom of fleeting, ghostly images."[32] This uncertainty is the fundamental ingredient of quantum theory. It leads directly to the consequence of unpredictability. "Most scientists...accept that the uncertainty is truly intrinsic to nature [and that] when it comes to atoms, the rules are those of roulette."[33]

Electrons and other particles simply pop about at random, seemingly without rhyme or reason. When we look for the location of an atom, we see an atom at a place. When we look for its motion, we get an atom with a speed. But we cannot fix on both factors at the same time. Thus, the world of atoms and subatomic particles is an uncertain and unpredictable world. Biologists who reject the divine connection to the origin of the species are like atoms living in their uncertain, chaotic universe. Atomic particles are incapable of imagining, based on their limited experience, that a predictable nature and a coherent universe can exist above all the uncertainties of their "colossal" micro-world.

Divergences between materialist biologists and the Quran occur not over proven facts but over inferences drawn from these facts. Biologists infer that gradual changes in life forms are due to the cumulative mutations dictated by environmental changes alone. They do not believe in the need to invoke a supernatural power to explain life and its origin. On the other hand, the Quran declares that God created all forms of life, with their deficiencies and adaptabilities, in the environments in which they exist. It maintains that the extinction of old species and the appearance of new ones are governed by divine laws and by the arrival of future moments packed with optional possibilities for God's creatures with free will to actualize their selected options into monuments of divine creations. Muslim scholars long before Darwin believed that environmental factors cause changes in life forms, but they also held that such factors are subject to God's guidance. For example, Ibn-Khaldun, after attributing the origin of human races to environmental factors, corroborated his point by quoting the Quranic verse: "This is how God proceeds with His servants. And verily, you will not be able to change God's ways."[34]

Another discrepancy between scientists and the Quranic teachings involves the assertion of the latter that the highest and most perfect form of God's creation occurred when the Divine Spirit (*Ruh*) breathed into animal bodies. The Quran explains the word *Ruh* in this way:

> **17:85** And they ask you about soul *[Ruh]*. Say: The soul is one of the commands of my Lord, and you are not given aught of knowledge but little.

The Quran states:

> **15:28–29** And when your Lord said to the Angels: Surely I am going to create and shape a mortal from the essence of black mud. When I have made him complete and breathed into him My Spirit [Ruh], fall down and make obeisance to him.

Thus, while materialists and the Quran share the concept of the evolutionary progression of earthly existence, they make different inferences as to the cause of such progression.

We have surveyed statements in the Quran that point us to the origin of humanity. Scientists and Muslims of our time might say that I am trying to reconcile the Quran with modern science, but data presented earlier confirms that many early Muslims believed in the origin of life as described in this chapter. Centuries before Darwin, Ibn Miskawayah, Ibn Khaldun, Ibn Arabi, al-Biruni, Ibn Sina, and other medieval Muslims taught that man has an ancestral link to the ape; yet such information was shocking news when Darwin conveyed it to the Judeo-Christian West. I do not need to fit the Quran into modern scientific beliefs because, conversely and unknowingly, modern science confirms the accuracy of the Quranic verses and what early Muslims wrote centuries ago.

Finally, the Quran reveals the concept of "creation by modification" rather than "the descent by modification," and the pre-Darwinian Muslim scientists concur with this concept of creation by modification. This modification occurs through mutations in the genes, which in our time frame, may have taken place over a period of four billion years, but in God's time frame, happens in no time and is a current event.

In summary, Muslims believe that the natural forces impacting life and vice versa are caused by divine laws, while modern scientists believe that these natural laws are the original and final answer to all of our questions, despite the fact that they do not explain how natural laws came into existence. Additionally, Muslims believe that although natural laws and events are comprehensible, God is incomprehensible. The Quran reads:

> **6:103** No vision can grasp Him, but His grasp is over all vision: He is above all comprehension, yet is acquainted with all things.

In our survey of early Muslim thought and the study of the Quran, we have learned that Muslim scholars described the evolutionary process in detail. Although they and the Quran did not use the terminology used by scientists today, they believed in the ever-growing bush called life long before modern scientists such as Stephen Jay Gould or Ernst Mayr. Not only did Western scholars hijack the fundamentals of the Muslim theory of evolution and present it as their own pristine idea, but they also excluded Muslims from their archives knowingly or unknowingly. For example, The Encyclopedia of Evolution by Richard Milner, with foreword by Gould, excludes Muslim authors from the history of the theory of evolution.[35] That omission is comparable to writing the history of the United States without mentioning George Washington, John Adams, Thomas Jefferson, or Abraham Lincoln.

CHAPTER 8

Adam and Eve: The Spiritual, Not Biological, Parents of Human Species

M OST MUSLIMS, if asked about Adam and Eve (*Hawwa*), will state that they were the very first of humankind on the face of the earth and that every human being of the past and current age is a genetic descendant of this primordial couple. The verse 4:1 is enlisted to support such a belief. But this belief crumbles under scientific as well as Quranic scrutiny.

A "genetic bottleneck" is a scientific term for a natural event that causes reduction in the genetic diversity of a living population. When a population is limited to a single sub-species, it lacks diversity and a significant percentage of that population fails to thrive due to disease, environmental challenges, and reproductive failures. Without a diverse gene pool, the viability of such a population gets reduced by 50% or greater. Eventually, this continuing deterioration of species vigor leads to its extinction. Small and isolated species populations are known to suffer dire consequences when a lack of genetic diversity exists. Among an isolated group of South African Muslims, the prevalence of a genetic disease, *cleidocranial dysostosis*, an inherited disorder of bone development, is traced to a common ancestor who was a religious and cultural leader of Chinese descent. Similarly, detrimental genetic diseases are very common in small interbred populations, and a genetic vulnerability of this nature coupled with a lack of genetic diversity can lead to the extinction of such a population. The smaller the population, the greater the probability it would disappear.

The smallest possible population of human species is two people, in a situation where naturally genetic variation is extremely limited. Therefore, the consideration that the proliferation of the human species could have resulted from only two ancestors such as Adam and Eve is not plausible by scientific criteria, a reflection from nature. The human species would have become extinct due to this extreme population bottleneck. So where, then, do Adam and Eve fit in the evolutionary history of our species?

Contradictory to commonly held view of Adam and Eve being the primeval parents of human species, the literal reading of the verse in association with other verses guides us to an alternate conclusion that Adam and Eve are really not the biological parents but the spiritual parents of human species. In order to determine the role played by Adam and Eve in the history of the human species, we need to have an understanding of the human soul-body relationship based on the Quran. A complex metaphor can be confused easily with reality. When we use pronouns such as "I" or "you," we recognize the person based upon the physical features of his or her material body. By doing so, we in fact are denying our own existence and accountability on the Day of Judgment. According to the Quran, the human soul is the spiritual center of the real human while the material body of flesh, blood, and brain serves only as a visible metaphor and a product of evolution. The material body is only the robotic arm of a soul.

Surgeons transplant hearts, lungs, and pancreas, kidneys, hands, and even faces onto the soul's outer garments of flesh and blood. Soon, surgeons will be transplanting laboratory-grown body parts into *Homo sapiens sapiens*. Still, each human soul maintains its individuality and acts in the material world as he or she did before surgery, if there is no damage to the brain. Therefore, if a physical body is taken out of the comprehensive human context, it is only a machine with component parts that can be modified, or if possible, replaced.

The dual nature of human body and soul is made very clear in the Quran as well as in the sayings of the Prophet Muhammad. According to the Quran, the souls are created and everlasting. It is alive and movable. "God takes souls at the time of their death and those who have not died in its sleep. Then He retains those for whom He ordained death and releases the others until a specific time. There are indeed signs in that for a people who reflect." (39:42) Hence, the material body can remain alive even when the soul is temporarily taken out from the sleeping body, thus showing our dual nature. Moreover, according to a sayings attributed to the Prophet, a fetus is alive but without a soul until the 120th day of intrauterine life, when the soul is breathed into a fetus by an angel.[1]

The verses 36:65 and 41:20–21 state that material body parts will testify against souls on the Day of Judgment: "This day [Judgment Day] We seal their mouths, and their hands speak to Us and their feet will bear witness concerning what they earned." (36:65) "Their ears and their eyes and their skins will testify against them as to what they used to do. And they say unto their skins: 'Why did you testify against us?' They say: 'God, who gave all things power of articulation, made us to speak.'" (41:20–21)

While the human spirit is accountable on the Day of Judgment, its material body is not. Therefore, the human spirit is separate from the physical body which is a product of evolution on the earth. However, on the Day of Judgment, humans will be raised up with all parts of their body as they were on the earth. The chapter, "The Resurrection," states: "Does man reckon that We shall not assemble his bones? Yea indeed, We are also able to shape his fingertips." (75:3–4)

Many Muslims use the Quranic verses 4:1 to discredit the theory of evolution as well. With the definition of humans as above, these verses definitely do not conflict with the theory. "O mankind, venerate your Lord, who created you from a single soul, from it created its mate, and from the pair of them has spread abroad countless men and women." (4:1) The creation of the human from a "single soul" and "from it…its mate (duality)" refers to the creation of the primordial human soul and its progression into a dyad in the generation of the male and female human souls, not yet clothed in the garment of biological flesh. Out of these two souls, God created all other human souls. What the verse refers to here is not the creation of visible metaphorical humans in flesh and blood but the creation of real humans manifested as spirits.

Next we will discuss the creation of the Prophet Adam. There are four steps in the creation of Adam, the prophet:

1. The creation of the soul of Adam,
2. The creation of Adam's outer garment, the physical body (*Homo sapiens sapiens*),
3. The infusion of Adam's soul, and
4. The elevation of Adam as the first prophet through revelations of divine messages to him.

These steps are described in the verse 15:29 and verse 2:31: "So, when I shaped him and breathed in to him my Spirit, fall down prostrating yourselves to him." (15:29) "Then He gave Adam knowledge of the nature and reality of all things and everything…" (2:31). These verses describe the creation of Adam's physical body and infusion of his soul into it.

Material Adam was an offspring with parents is suggested by the following verse: "God did choose Adam and Noah, the family of Abraham, and the family of Imran for high distinction. The one is the progeny of the other God heareth, knoweth." (3:33–34) The phrase "offspring, one of the other (*zurriy-yatam-ba'-zuhaa mim-ba'-z*)" in verse 3:34 is binding on Adam, as the context clearly shows, which supports the theory that Adam and Eve were the offspring (*zurriyah*) and parents.

Ordinary humans become prophets only when God reveals His wisdom to them. So, God elevated Adam to become the first of His long chain of prophets by teaching him "knowledge of the nature and reality of all things and everything." (2:31). Only then, God elevated *Homo sapiens sapiens* above non-physical beings such as angels and *jinns* and asked them to pay homage to Prophet Adam. God blessed him to live in a spiritual "Garden" of ease, happiness, and innocence.

Muslims believe that Islam is the religion of all of the prophets. This belief is based upon the following verses: "Truly the Religion with God is Islam..." (3:19), "Say: We believe in God, and that which has been sent down to us, and that which has been send down to Abraham, and Ismael, and Isaac, and Jacob, and the tribes, and that which has given to Moses and Jesus, and that which the prophets received from their Lord. We make no distinction between any of them, we surrender to Him," (2:136) "Abraham was neither a Jew, nor yet a Christian. But he was a hanif, a Muslim..." (3:67). According to these verses, all prophets preached Islam, and so they are all Muslims. Adam being the first prophet of God preaching Islam, he and his mate became the original spiritual parents of all humans.

If Adam and Eve were the primordial parents of human species, incest was practiced contradictory to Islamic morality. The divine religion of Islam preached by all prophets does not permit marriage between sisters and brothers. Therefore, Adam and Eve were not the biological parents but the spiritual parents of the human species. If Adam was the first prophet, he must have had a human community to whom to preach the divine message. Let us explore whether there is any evidence for the existence of a community of human species as contemporaries of Adam.

In Chapter 7, we have already learned about God-directed evolutionary process that led to the creation of the visible outer DNA-based garments of real humans, named by modern biologists as *Homo sapiens sapiens*. Then verses 7:10–11 describe the creation and the settlement of the initial human community on the earth before the creation of the Prophet Adam. The verses state: "And We have given you (mankind) power in the earth, and appointed for you therein livelihoods. Little give ye thanks! And We created you, then fashioned you, then (ثم) told the angels: Fall ye prostrate before Adam! And they fell prostrate, all save Iblis, who was not of those who make prostration." (7:10–11)

A close examination of the grammatical structure of the verses remains very important. In many languages, verbs are conjugated in relation to the number of individuals to which the action applies. In English, every noun is either singular or plural. Every verb is modified to comply with a singular or plural noun or pronoun. Unlike English, Arabic has singular, dual, and plural forms of pronouns, nouns, verbs, and adjectives. The singular form is used when referring to one person or thing, the dual refers to two people or two things, and the plural form for more than two people or things.

Certainly .... وَلَقَدْ

We have established you [humankind]
(more than three plural) .... مَكَّنَّاكُم

in the earth .... فِي الْأَرْضِ

and made for you (more than two plural) .... وَجَعَلْنَا لَكُمْ

in it .... فِيهَا

livelihood .... مَعَايِشَ

And certainly .... وَلَقَدْ

We Created you (more than two plural) .... خَلَقْنَاكُمْ

Then .... ثُمَّ

Fashioned you (more than two plural) .... صَوَّرْنَاكُمْ

then .... ثُمَّ

told the Angels: fall ye prostrate
before Adam! .... قُلْنَا لِلْمَلَائِكَةِ اسْجُدُوا لِآدَم

The grammatical construction of the verses with the plural in مَكَّنَّاكُم (established you), وَجَعَلْنَا لَكُمْ (made in it), خَلَقْنَاكُمْ (created you), and صَوَّرْنَاكُمْ (fashioned you) is the confirmation that the initial part of the verse is not speaking of the creation and the settlement of Adam on the earth, but the creation and settlement of the initial human community. God blessed the original community of human species with Prophet Adam as their guide, and only "**then** (ثُمَّ)" did God ask angels and *jinn* to prostrate themselves before Adam. There are additional Quranic reasons to believe in the existence of human contemporaries with Adam.

The following Quranic verses 2:35–36 narrate the account of the expulsion of Adam and Eve from the "Garden."

And We said:	وَقُلْنَا
O! Adam (singular)	يَا آدَمُ
You dwell (singular)	اسْكُنْ
You (singular)	أنتَ
and your mate (singular)	وَزَوْجُكَ
in the garden	الْجَنَّةَ
and you two eat (dual)	وَكُلَا
from it	مِنْهَا
eat freely as	رَغَداً حَيْثُ
you two wish (dual)	شِئْتُمَا
and you two approach not (dual)	وَلَا تَقْرَبَا
this tree	هَذِهِ الشَّجَرَةَ
lest you two become (dual)	فَتَكُونَا
transgressors.	مِنَ الظَّالِمِينَ
Then Satan made them to slip (dual)	فَأَزَلَّهُمَا الشَّيْطَانُ
from it	عَنْهَا
and send the two out (dual)	فَأَخْرَجَهُمَا
from that (State) the two (dual)	مِمَّا آلَانَ
were in	فِيهِ
And We (God) said:	وَقُلْنَا
'Get you down. (more than two plural)	اهْبِطُوا
Some of you (more than two plural)	بَعْضُكُم
are enemies to others; (more than two plural)	لِبَعْضٍ عَدُوٌّ
and you have (more than two plural)	وَلَكُم
in the earth a habitation and sustenance for a while	فِي الْأَرْضِ مُسْتَقَرٌّ وَمَتَاعٌ إِلَى حِينٍ

Verse 2:35 begins with a singular noun and pronoun, indicating Adam, then changes to a dual pronoun referring to both Adam and Eve. The Arabic verb used in the Divine command to get out (اهْبِطُوا) of the "Garden" in verse 36 after the violation of the Divine guidance, is conjugated, however, not with the dual form, but with the extended plural indicating more than two subjects. The medieval

Muslim scholars have interpreted this in two different ways. One is that the command in plural, "get you down" (اِهْبِطُوا), is directed to a group consisting of Adam, Eve, and Satan. The second proposes that the command is given to a group consisting of Adam, Eve, and their future, yet-to-be born descendants. Both of these interpretations invariably lead us to a genetically unviable bottleneck that would have predisposed early humanity to extinction.

As the interpretations of the early Muslim scholars do not find credence within a realistic biological model, I offer a third interpretation. This interpretation is compatible with the modern scientific understanding. Returning to the verses 2:35–36, initially Adam and Eve are addressed while in the Garden as "you" in the dual grammatical form when God offered them to enjoy the "Garden." Adam was nevertheless deemed a Prophet. Adam must have had contemporary human beings by whom he was ordained to convey the Divine message. God ordered them to "get you down" (اِهْبِطُوا) from their paradisiacal mental state of perfect ease and happiness when the Prophet Adam and his congregation, the trustees of God on the earth, violated the divine guidance. The use of the plural form, اِهْبِطُوا ("get you down"), in the divine command suggests the presence of a generation of human beings along with Adam and Eve. Then, the Quran states, "Some of you (بَعْضُكُمْ) are enemies to others." The use of the plural form, بَعْضُكُمْ (some of you), referring to a group of human beings is another reason to suppose that there were other human beings with Adam and Eve. So, based on Quranic verses 2:35–36 and 7:10–11, God peopled the world of Adam and Eve with rich genetic diversity so that His vicegerents would not go extinct due to the population bottleneck until the end of the world.

CHAPTER 9

The Location of the Garden of the Forbidden Tree

MOST CONTEMPORARY MUSLIMS across the world believe that Adam and Eve were created in Paradise *(Jennat-ul-Khuld)* but were expelled for eating fruit from the forbidden tree in the garden. Early Muslims carried on great debates about the location of the garden. According to the two foremost exegetes of the Quran, Ibn Kathir (died in 1372) and ar-Razi (died in 1209), four interpretations of the location of the garden prevailed: that the garden was Paradise itself, that it was a separate garden created especially for Adam and Eve, that it was located on Earth, and the view that it was best for Muslims not to be concerned with the location of the garden.[1 & 2]

Unorthodox as it seems for our time, more reasons lead us to believe that the garden was on Earth rather than in Paradise. Many Quranic verses suggest that Paradise is a place the eyes of believers have never seen:

> **32:17** No living soul knows what comfort of eyes has been kept hidden from them as a recompense for their deeds.

Abul A'la Maududi, a twentieth-century commentator of the Quran from the Wahhabi faction among Muslims, tells us that Bukhari, Muslim, Tirmidhi, and Imam Ahmad, all most respected scholars in the field of the Prophet's traditions, cited Abu Hurairah's text (a contemporary of the Prophet) that the Holy Prophet said: "Allah says: I have prepared for my righteous servants that which has neither been seen by eyes, nor heard by ears, nor ever conceived by any man."[3] If no human eyes have seen it, then Adam could not have seen it, and therefore the Garden of the Forbidden Tree could not have been in Paradise.

The contextual analysis of the following verses also rules out the possibility of the garden being Paradise:

> **2:31–34** And He taught Adam the nature of all things; then He placed him before the Angels, and said: Tell me the nature of these if you are right... And behold! We said to the angels: bow down to Adam and they bowed down. No, said Iblis (Satan). He refused and was haughty. He was of those who rejected the Faith.

When Iblis refused to prostrate before Adam, God ordered him out of the garden. God said:

> **7:13** "Get thee down from this: it is not for thee to be arrogant here: get out, for thou art of the meanest (of creatures)."

Then in the presence of Adam and Eve before their expulsion from the Garden, Iblis requested and God replied:

> **7:14–15** "Give me respite till the Day when they shall all be raised up from the dead." Allah replied, "You are granted respite."

According to the above verses, Adam learned the nature and reality of all things and everything (2:31). Therefore, when God granted Iblis respite, Adam and Eve must have known that their lives in the garden with the forbidden tree were subject to imminent death and that they would be "raised up from death" on the Day of Judgment (7:14–15). For these reasons, the garden could not have been in

Paradise and it must have been on earth. The promise of the Garden of Final Return would make no sense for Adam, as well as for his human contemporaries for their good deeds, if they were already there.

Iblis deceived Prophet Adam and his followers, telling them: "Your Sustainer has forbidden you this tree, lest you two become [as] angels, or lest you live forever." (7:20) He whispered to Adam: "O Adam! Shall I lead you to the tree of eternity and to a kingdom that never decays?" (20:120) Thus, by describing the Forbidden Tree as a secret recipe for eternal life, Iblis tempted Adam and Eve to partake of the fruit of the tree. When they consumed the Forbidden Fruit, they sought eternal life. The Quran states that the Garden of Final Return is eternal for the righteous and the Hell "for whoever of them follows you, indeed the hell shall be your requital, an ample reward."(17:63).

If Adam and Eve were actually in it, then they too would be eternal. The desire to acquire a thing develops in the human mind only when the desired object or state is not available. Adam was in an environment where death and decay, not immortality, were the facts of life. Adam and his contemporaries would not have become victims of temptation if they had already been immortal beings in Paradise. So, the Forbidden Tree as well as Adam, Eve, and their human contemporaries were not in the Garden of Final Return.

The following verse adds more support to the falsity of the notion that the forbidden tree was in Paradise.

> **37:7–10** I have secured it (Heaven) against every rebel Satans. These Satans cannot hear the words of the exalted ones; they are darted at and driven off from every side, for them there is a perpetual torment. However, if someone snatches away something, a flashing flame follows him.

Abul A'la Maududi, a scholar among the Wahhabi faction of Muslims, explains the above verse as follows: "The satans have no access to Heaven and no power to hear the angels' conversations…if by chance a little of its news enters the ear of a devil, and he tries to bring it down, he will be followed by a flashing flame."[4] Based on verses 20:120, 7:20, 37:7–10, and 7:22, we can conclude the following: First, Adam was allowed to stay in the Garden of the Forbidden Tree after Iblis' expulsion from it. Second, although Iblis did not have access to the Garden of Final Return, he succeeded in entering and deceiving Adam. Third, Adam and Eve fell for Satan's persuasion and ate from the tree in the hope of becoming immortal. Thus, the Garden of the Forbidden Tree was a place where all that existed had a transient, decaying nature into which Satan was able to enter. How could it be the Garden of Eternity?

Quranic verses repeatedly disprove the idea that the garden was in Paradise. One verse reads: "There [Paradise] they call for every kind offruit in peace and security: Nor will they there taste death...." (44:55–56) Thus, the presence of a forbidden tree bearing the fruit of immortality and the commandment that Adam and Eve not eat of that fruit contradict the location of the Garden of Forbidden Tree in Paradise.

Throughout the Quran, Paradise is described as a place where peace prevails: "There they shall hear no idle talk, but only 'Peace'" (19:62), where sin, lies, and vanity do not exist (56:25–26; 78:35, and 88:11). If we accept the Quranic description of the Garden of Final Return, then we also have to conclude that the Garden of the Forbidden Tree could not be the place where Satan makes the false, vain claim that he is better than Adam: He (Iblis) said:

> **7:12** "I am better than he [Adam]; You have created me of fire and created him [Adam] of clay."

If Iblis' claim were true, God would not have asked the angels and the *Jinn* to prostrate before Adam—thus Iblis lied. In Paradise, Iblis would not have been able to tempt Adam to commit sin.

Such egoism, lying, and deception could not have taken place in In the Garden of Final Return—a place where fruits are not forbidden, where peace and justice prevail, and where sin, lies, and vanity are unknown.

Finally, Adam could not have been thrown out of Paradise because it is the place of final return: "... and verily, for the righteous, a beautiful place of (final) return." (38:49) Thus, Adam, Eve, Iblis, and the Forbidden Tree could not have been in Paradise (*Jennat-ul-Khuld*), and the notion that the garden with the forbidden tree and Paradise were one and the same is false. The Garden of the Forbidden Tree is a place for human trial and error, while Paradise is a place for reward.

Since the Quran does not support the notion that the garden was Paradise, where was the garden instead? The following verses offer clues. After God presented Adam to the Angels, He said to Adam:

> **20:118–119** There is therein (enough provision) for thee not to go hungry nor to go naked, nor to suffer from thirst, nor from the sun's heat (تَضْحَ).

God provided Adam and Eve with food, water, and shelter to protect them from the heat of the sun. From among the billions of stars in our universe, why would the Quran mention the sun's heat if the garden of the Forbidden Tree was Paradise? Moreover, in every context (7:98, 20:59, 79:29 and 46, and 91:1), the word (daha ضحى) or its derivatives are used in the Quran only in situations related to sun. Thus, the Garden of the Forbidden Tree must have been within the solar system in order for it to be affected by the heat of the sun, which could be felt only on planets close to it. The Garden of the Forbidden Tree could not have been on Mars, Jupiter, Saturn, Uranus, Neptune, or Pluto because their surface temperatures are so low that mention of the protection from the sun's heat would be illogical. The surface temperatures of Mercury and Venus are extremely high, too high for plants and animals to exist as living creatures and as sources of food for Adam. The only place in the solar system where DNA-based life exists and where humans need protection from the sun is Earth. Based on our present-day knowledge of the solar system and on the above verses, we can conclude that Adam and Eve were in a garden that was located on Earth. (Figure 9-1).

The Quran tells us also that God created Adam and Eve on the earth with earthly material and granted a fixed term of life there: "Behold, thy Lord said to the Angels: I will create a vicegerent (khalifah) on earth" (2:30) and "It is He who created you from clay, and then decreed a stated term [of life] for you...." (6:2) Moreover, we have seen in Chapter 8 that human species were settled down on the earth before the presentation of the Prophet Adam to angels and

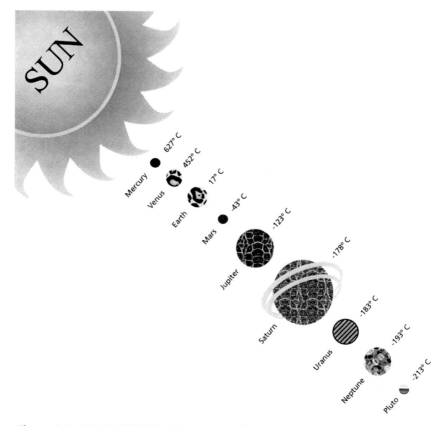

Figure 9-1. SOLAR SYSTEM. Relative sizes of the planets arranged in order of distance from the sun and the surface temperatures of the planets.

Jinns and subsequent deception of Adam by *Iblis,* confirming that the Forbidden Tree was on the earth:

> **7:10–11** And We have given you (mankind) power in the earth, and appointed for you therein livelihoods. Little give ye thanks! And We created you, then fashioned you, then (ثُمَّ) told the angels: Fall ye prostrate before Adam! And they fell prostrate, all save Iblis, who was not of those who make prostration.

Finally, the name bestowed by the Quran on the first perfect human being is highly significant. The name "Adam" is derived from two Arabic root words, *adim* and *udama*. *Adim* means surface of the earth, and *udama* means dark-complexioned.[5] The literal meaning of Adam, therefore, is a dark-skinned person who lives on the surface of the earth. This, too, supports the idea that Adam was created on earth and not in Paradise, and was of African origin.

As previously emphasized, the Quran asks human beings to observe and discover the mysteries of life on Earth as well as in the universe:

> **51: 20–21** On the earth are signs for those who have faith (in the meaningfulness of all things), as also (there are signs) in your own selves: Will you not, then, observe?

What are those mysteries, and what should we observe? The Quran urges us to trace and explore the human evolutionary journey of our primordial parents, Adam and Eve. Thus, we must remain aware of important scientific facts. First, protoplasm is much the same in all organisms, suggesting a unity among all of them. Similar physiological pathways catalyzed by similar enzymes occur throughout all life.

At the molecular level of all life forms, we find homologous series of chemical molecules and compounds. The degree of differentiation in the sequence of molecules in the homologous compounds is roughly proportional to the degree of taxonomic separation (i.e., classification of animals or plants according to their natural relationship) of the species in which the compounds occur. We also need to realize that fossil records document the progression of life through time, and that we can change the shape and function of life forms by changing the genes. We observe that hemoglobin in the red blood cells has two chains of amino acids—alpha and beta chains. Closely related organisms differ less in the amino acid sequences of their hemoglobin alpha chains than do distantly related organisms. The same is true when beta chains are compared with each other. There are many other human observations in nature that I could list here as well.

The above observations and many more suggest that all living forms are related to each other. The grass and the grasshopper and the monkey and humankind can be sisters, brothers, cousins, grandparents, and grandchildren. They all developed from a single seed called DNA, through genetic modification.

Many Muslims who believe that Adam and Eve were created in Paradise promote their view by quoting verse 2:38. According to the verse, when Adam disobeyed God's command and ate the Forbidden Fruit, God ordered him out of the garden, saying: "Get ye down all from here." (2:38.) They claim that the verse suggests that Adam went from a higher place to a lower place. They believe that Paradise is located above the earth. Such belief is problematic and unacceptable; we know that the earth rotates on its axis and that it is an integral part of the Milky Way Galaxy which also rotates on its axis. Therefore, if Paradise were visible from Earth, we would see it beneath us at one point in time, above us at another, and at various times on either side. How, then, can we explain the above verse? Many early Muslim commentators believed that, before their expulsion, Adam and Eve experienced a mental state filled with ease, happiness, and innocence and that the phrase "get ye down" shows they were driven out of a blissful mental state. Muhammad Rashid Rida records that explanation in *Tafsir-ul-Manar*.[6] Regardless of one's concepts regarding the location of Paradise, based on the above study of the literal reading of the Quranic verses, one can reasonably conclude that Adam, Eve, and the forbidden tree were located on the earth and not in the Garden of Eternity.

CHAPTER 10

God and Chance Events

C HANCE OR RANDOM EVENTS are an integral part
of the theory of evolution. What do we mean by
chance? The simple meaning of the word is some-
thing that happens unpredictably without discernible human intention or direction, and in dissociation from any observable pattern,
causal relation, or natural necessity. We define events as unpredictable in two ways. In the first, we are limited by the lack of detailed
and accurate knowledge at the micro-level. In such cases, *chance*
means that we are unable to determine all the micro-factors affecting
the initial conditions that determine the macro-events. In the second scenario, the observed "unpredictable" event is the result of the
crossing of two or more independent causal trains of events, when
we have had no precise information about the trains themselves or
about their point of intersection, or about the origins of preceding
small, sequential, causal events.

Many Muslims believe that assigning a role for chance in the creation of the universe and its subunits—animate and inanimate—is
anti-Islamic, a denial of God's absolute control of the universe. In
their view, the universe is like a clock and God is the clockmaker.
The clockmaker set the clock in motion to tick away toward the

Day of Judgment. This miraculous clock never slows down, never runs fast, and never stops. A perfect clock! But if the universe does run like a clock, then future events should be precisely predictable. Our whole lives and futures would be predetermined, and we would have no freedom to actualize any possibility or affect any change in the universe. If the universe is a mechanical clock whose function was predetermined in the past, all God could do is helplessly observe it ticking away to the Big Crunch. We would have an unemployed God!

A universe without chance and a variety of possibilities exists only if creatures do not have the potential or freedom to help or to hurt, to believe or not to believe in God, and so on. In a deterministic, chance-absent universe, humans cannot choose freely what they want, but are forced to submit to destiny, and God is responsible for unthinkable evils.

In such a universe, for example, Wahshi, a hired killer, had no choice but to kill Hamzah, the uncle of the Prophet, in the Uhud war, because it was meant to be. Hind bint Utba, who paid Wahshi to kill Hamzah and who then cannibalized him, also had no choice in the matter because it was destined to happen. In such a fatalistic universe, Osama bin Laden and other fanatics among a minority of misguided Muslims played no role in forming American public perception of Muslims. Instead, all is destiny!

God becomes the ultimate puppeteer who pulls the strings on puppets such as Wahshi and Hind bint Utba to kill and mutilate one of the companions of the Prophet (pbuh), and God becomes responsible for hiding the true face of Islam from ordinary Americans so that they see all Muslims as terrorists. In a deterministic, chance-absent universe, human choices and actions are useless because destiny, not chance or volition, transforms them into believers or nonbelievers!

We can discover many other reasons to reject the deterministic, chance-absent universe of Muslim anti-evolutionists. According to the Quran, the current universe is a testing ground for us to gather data in preparation for the Day of Judgment, when God will either reward or punish us based upon His evaluation of what we gathered during our lifetime. The Quran states: "[Allah] Who created death and life in order to try you to see who of you are best in deed" (67:2) and "Or do ye think that you shall enter the Garden (of Bliss) without such (trials) as came to who passed away before you?" (2:214) It also reads: "On no soul does God place a burden greater than it can bear. It will receive every good it earned and suffer every ill it earned...." (2:286). Therefore, in the Islamic universe, every human being is a free agent given equal chances and free will to perform moral or immoral acts—equal chances to get to Heaven or Hell.

What each human being does on this earth matters on the Day of Judgment, when the Divine's absolute justice prevails.

In the predestined universe posited by anti-evolutionists, our thoughts and resulting deeds are preordained by the clockmaker. But if this were so—if Allah prompts every move of His earthly creatures—genuine tests or trials as described in the Quran will lead to accountability for deeds over which we have no control? The Day of Judgment becomes a tyrant's phony court hearing where pre-programmed robots called humans, who have no control over their actions or decisions, are capriciously judged. Without genuine freedom and the chance to do good or evil, where can humankind find the Generous (Al-Karim), Merciful (Ar-Rahman) God?

Many medieval Islamic scholars and imams have rejected a deterministic universe. For example, in response to a letter from Caliph Abd al-Malik ibn Marwan regarding the doctrine of predestination, Imam al-Hasan al-Basri (b. 624) replied:

> O Commander of the faithful; do not alter it or interpret it falsely. God would not openly prohibit people from something and destine them to do it secretly as the ignorant and the heedless say. If that were so, He would not have said in the Quran, 41:40: *Do what you wish* but would have said: 'Do what I have destined you to do'. Nor would He have said in Quran, 18:29: *Whoever wills shall believe and whoever wills shall disbelieve* but would have said: 'Whoever I will shall believe and whoever I will shall disbelieve'.[1]

The fundamental purpose of the Quran and the Prophet's authentic traditions is to make us aware of Heaven and Hell and of the moral or immoral decisions that we could make by use of our free will in an unpredictable world. These sources are also decision-making guides as to what actions God allows, prohibits, or considers neutral. God does not, however, totally determine or preordain individual choices and actions. In this universe, we are free to choose from contrasting possibilities that are packed as information in each moment of the arriving future. We are free to actualize our choices into visible monuments of Allah's creation in the material world.

God holds us accountable if our choices violate the guidance that He revealed though his prophets. The All-knowing God (al-Alim) knows that free will exists only where authentic choices exist. So, He offers us an open future with chance events and a multitude of possibilities that allow us to exercise our free will. Human beings, unlike robots, are even allowed to accept or to reject God, because they are not preprogrammed. Allah is not a puppeteer. He is the All Compassionate, Merciful, and Just God who presents us with possibilities and who permits us the freedom to choose right from wrong or vice versa. Thus, we see that random and chance events inevitably occur in this universe peopled with creatures granted free will.

Anti-evolutionist Muslims acknowledge the truth that Allah used the elements of chance and unpredictability in the historical processes that produced our contemporary universe. Anti-evolutionist Muslims cannot deny this fact, because the revelation of many Quranic verses emerged as a direct answer to the free human choices of the Prophet and his contemporaries. A marvelous example is the revelation of chapter 80, "He Frowned (Abasa)." One day, the Prophet (peace be upon him) was engrossed in a conversation with some of the most influential chieftains of pagan Mecca, hoping to convince them of the divine message. At that point, a blind ordinary man approached him with the request for an elucidation of certain earlier verses of the Quran. Annoyed by this interruption of what he momentarily regarded as a more important endeavor, Muhammad (peace be upon him) "frowned and turned away" from the blind man. Immediately, there and then, the first ten verses of the chapter 80 (He Frowned) were revealed to reprove the Prophet's treatment of the blind man.[2] If chance existed in the daily affairs of human beings and even of the prophet, which shaped their histories and epics and even in their religions and holy books including the Quran, why should Muslims reject the role of chance in the shaping of our biological history? We can discern no reason to do so in the matters related to the evolution of the universe.

Chance is an integral, unavoidable part of the evolution of life. How did God design and direct this miraculous universe while allowing for chance and random events? What are its mechanics? How does God respond to and answer prayers without violating natural laws? How does He maintain the continuity and directionality in the evolution of the universe and life in the presence of chance? Let us explore these questions in the next chapter.

CHAPTER 11

The Structure of the Universe: A Master Design

MATERIALISTS SAY that the universe is not intelligently designed. Herein, however, we explore the structure of the universe as a product of an intelligent master design to serve the purpose and intent of God in the presence of free will in His creatures. Without understanding the structure of the universe, we cannot understand creation intelligibly. We begin, therefore, with a review of the basic structure of the universe.

As mentioned earlier, materialistic science claims that the future is totally predetermined by earlier causes and is predictable through accurate knowledge of the past because physical and chemical laws are invariant. Consequently, any event in nature that departs from an anticipated outcome of a known cause is considered coincidental.

Materialists, especially some molecular biologists, arrogantly state that all aspects of life will ultimately be explained in mechanical terms through knowledge of past events as recorded in geology, paleobiology, vestigial organs, and so on. They conveniently forget a particular past event that shocked materialistic mechanical science—the quantum theory of Max Planck and its modification by Niels Bohr. Even Bohr was overwhelmed and said, "Anyone who is not shocked by quantum theory has not understood it."[1] In the hands of Niels Bohr, the theory melted solid matter—the basic building block of the universe and its contents (including human)—into nonsolid energy. Twentieth-century relativity and quantum mechanics overturned and swept aside the concept of a predetermined universe and replaced it with an indeterminate universe.

According to the quantum theory, the behavior of matter is unpredictable. No one can say, for example, that a particle was in a definite position in the past or that it will occupy a definite place at some time in the future.

> "A particle such as an electron does not appear to follow a meaningful, well-defined trajectory at all. One moment it is found here, the next there. Not only electrons, but also all known subatomic particles—even whole atoms—cannot be pinned down to a specific motion."[2]

The more accurately we measure the momentum of a particle, the less we can calculate the position of a particle. The more we know about the particle's position, the less we can say about its momentum. Our partial information about the position of a particle only yields the probability that it is within a certain distance of a particular point. The probability factor of quantum mechanics presents itself in other ways. For example, when an atom collides with a photon of light, we cannot precisely predict what will happen, but can only speak of the probability that the energy of the photon will or will not be absorbed by the atom. Similarly, it is not possible to predict when a radioactive atom will decay. In a uranium (^{238}U) sample, the sudden decay of a specific atom into Thorium (^{234}Th) through the emission of alpha particles can only be computed as a probable occurrence. We are unable to explain why a particular uranium sample decayed while the uranium next to it with absolutely identical atoms did not. We may calculate a certain chance that the decay will take place within the next ten seconds. The probability exists that the remaining atoms may decay over a period of ten thousand years. But no one can give a definite answer regarding the sequence and time of decay.[3]

Thus, we see that the physical sciences cannot tell us when an event will take place, and sometimes it cannot even predict what will happen. If science cannot tell us what is occurring among matter—the basic building block of the universe and living organisms—how can it precisely predict the future of human beings or anything else in the universe? This scientific paradox bothered even Einstein, who nonetheless concluded, "God does not play dice."[4]

The quantum theory gave birth to new chemistry that explained how atoms are bonded together to form molecules. When two atoms initially separated are brought together in a chemical reaction, the electrons in their outermost shell (electrons the farthest from the nucleus) share one orbit. Such sharing in a chemical reaction is called a covalent bond (Figure 11-1).

In covalent bonding, atoms share electrons to form all molecules, including ordinary substances such as water, methane, and so forth.

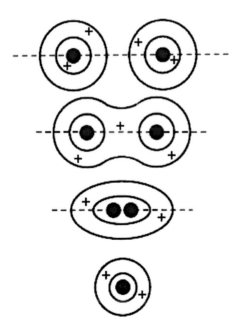

Figure 11-1. FORMATION OF COVALENT BOND

In some cases covalent bonding can lead to the formation of huge, extended macromolecules such as polymers. One example of such a polymer is deoxyribonucleic acid (DNA), the basic building block of life.

The formation and maintenance of the structure of compounds such as DNA is ultimately the result of quantum mechanics, in which the behavior of atoms remains predictable only as a statistical average. Every thought and every motion within all life forms remains inseparably connected to electricity and chemistry, and ultimately to unpredictable quantum events as well. If the behavior of matter and the formation of compounds are not absolutely predictable, can the future of any creature be precisely predictable? The reply of quantum physics is a resounding "no," because every DNA-based creature is only a mosaic made of atoms and chemicals. The best calculations that modern science can calculate are only the averaged probabilities of thousands of quantum events. In the quantum world, science loses its confidence when up against the law of causality. Hence, materialists' extremism against God gets exposed as unscientific at the quantum level.[5]

Unlike materialist scientists who proclaim that there is no purpose for the universe, the world's major religions assert that God created the universe by design and for a purpose. The Prophet Muhammad explains it as follows: "Allah said: I was a hidden treasure. I wanted to be known, so I created the world."[6] Allah did not want to place all of His creations in the same spiritual reality. Therefore, He created a separate universe. Allah made the larger components of the universe comprehensible to the human mind, the best of His creations, along with the installation of stubborn, immutable chemical and physical laws to function in a uniform and repetitive fashion at the level of those larger components. God did not intend the same immutable laws of classical physics and chemistry to work like a programmed machine all the way down to the atomic and subatomic levels. If the atomic world functioned in a uniform and predictable fashion, then we would be able to foresee the future accurately through our knowledge of past causes. If this were the case, humankind would lose their freedom to make moral choices. Allah would then be restricted to the role of a passive spectator, unable to act without suspending the laws of the universe in a way that is visibly obvious to humans. Therefore, God created matter as the building block of the universe and governed its actions through an immutable system of physical and chemical laws within which it functions in unpredictable ways. By doing so, He grants human beings the freedom to make moral choices about their future.[7] Due to the unpredictability of the structure of matter, absolute knowledge of the nature and future of our universe will never become known to us until the Day of Judgment.

Materialist scientists are unable to understand that God is an active participant in the world. Chemists have discovered that all components of the physical universe, including genes, are made of atoms arranged in different fashions. Genes are made out of DNA. DNA is a collection of nucleotides. Phosphate, sugar, and four amino acids (thymine, cytosine, adenine, and guanine) are chemically connected with covalent chemical bonds to form nucleotides. A genetic mutation can produce a major effect on the outward physical features of an organism. Mutations depend on the changes in individual molecules due to the breaking of specific atomic covalent bonds that involve quantum mechanical processes. Physicist-theologian Robert J. Russell points out that "this is ultimately a quantum process at the atomic level initiated by the breaking of a single hydrogen bond."[8]

When God prompts a small quantum fluctuation (jump of a quantum of energy) in the atomic or subatomic world and makes or breaks a covalent chemical bond or bonds of a gene, materialists see a mutation or an accidental birth of a species. They interpret it as a random or accidental event with no cause. They do not realize that God built life around chemistry that provides "the amplifying mechanism for quantum events."[9] Physicist William Pollard feels that if chemistry is the physical appearance of an organism (phenotype), the quantum fluctuation is the cause (genotype).

This construction of the understandable chemistry coupled with the unpredictable quantum physics is an ingenious intelligent design of the All-Knowing and All-Powerful Allah. Within this master design of the universe, Allah, by voluntarily limiting His omnipotence and omniscience, gave His creatures a genuine freedom and an open future while humans can still derive sensible meanings out of the complex universe at the same time. In such a material universe,

CHAPTER 12

Self and Subjectivity of the Animate and Inanimate World

MATERIALISTS RECRUIT genetic mutation, unexplained events, and pain and suffering in the material world as "proof" to reject God and discredit believers. Within the Islamic metaphysics of the future, however, these natural events are not contradictory evidences against God or against the intelligently designed universe, but complimentary glorification of God.

As we have seen, the creatures in the Islamic universe which include quarks and atoms as well as collections of atoms such as the stars, Earth, mountains, human beings, amoebas, etc. are constantly addressed through divine acts that present them with future possibilities in the form of information.

God endowed all creatures with minds. Animate and even inanimate creatures, such as fire, winds, and mountains have selves and a subjective faculty with which they can experience and respond to the Divine Will. The Quran attests to the existence of subjectivity with faculties endowing experience and response within the inanimate world, even though human beings do not comprehend it:

181

17:44 The seven skies, the earth, and all that lies within them, sing hallelujahs to Him. And there is nothing that does not chant His praise. But you [human] do not understand their hymns of praise. Truly He is verily clement and forgiving.

We also read:

41:11 Then he turned to the heavens, and it was smoke. So he said to the earth and the heavens: "Come with willingness and obedience and they replied: We come Willingly."

The latter verse can be interpreted scientifically. According to contemporary cosmology, the entire universe was filled with radiation and plenum of matter (originally hydrogen and helium) formed from elementary particles (quarks) in a dense, primeval fireball of creation called the Big Bang. The smoke described in the above verse is most likely a reference to quarks and atoms before they condensed into galaxies. The Quranic verses here describe a universe that was responsive to God since its inception after the Big Bang. The heavens and the earth in its early gaseous embryonic state ("smoke") responded to God by saying, "We come willingly."

Moreover, the Quran substantiates that natural events such as thunder, fire, and wind have self and subjectivity: "And the thunder extols His praise, and the angels are in of awe of Him...." (13:13). The subjectivity of fire is well documented in other verses also. When Abraham was cast into fire, Allah said: "...O fire, be you a coolness and a safety for Abraham." (21:69). A verse relating to Solomon reads: "So We subjected the wind to him [Solomon]; it ran softly at his command to wherever he pleased." (38:36) These verses guided Jalaluddin Rumi to write: "Air and earth and water and fire are (His) slaves. With you and me they are dead, but with thy God they are alive."[1]

According to the Quran, Allah offered the heavens, earth, and mountains the opportunity to bear the responsibility of carrying out His will, giving them the choice to deviate from it at the risk of punishment. They refused because they were afraid of violation and consequent punishment. The verse reads: Surely We presented the trust to the heavens and the earth, and the mountains, but they declined to bear it and they feared it, and man bore it, but he is iniquitous, ignorant. (33:72) Ibn Kathir (1160–1234), an important medieval exegete of the Quran, noted many Traditions of the Prophet from that of Abdullah Ibn Abbas, a companion of the Prophet, to that of Hassan al-Basari (d.728), which reflect the subjectivity of the inanimate world.[2] Similarly, Qurtubi (d.1273), another well-respected exegete, using the authoritative text of Tirmithi (d. 892), who had quoted Abdullah Ibn Abbas, wrote that after the refusal of the heavens, the earth, and the mountains, Allah offered this trust to Adam, who accepted it.[3] Thus, the Islamic faith and its traditions confirm the subjectivity of the inanimate world.

In the spiritual universe of Islam, even a part of a whole organism has its own inherent individual subjective faculty with which to experience and to respond to God and has a separate physical existence from the whole, as we can see from the following verse:

> 36:65 This day [Judgment Day] We seal their mouths, and their hands speak to Us and their feet will bear witness concerning what they earned.

Another verse warns that on the Day of Judgment:

> **41:20–21** …their ears and their eyes and their skins will testify against them as to what they used to do. And they say unto their skins: Why ye testify against us? They say: God hath given us speech, who giveth speech to all things, and who created you at the first, onto whom ye are returned.

Thus, we see that the individual components of organisms, such as their ears, feet, skin, DNA, genes, etc. have some measure of feeling that will allow them to respond to the guidance of God.

The inner potential in lower orders of creatures may be minimal and hard to detect. As God created higher orders of being, He carefully gifted each of them with more potential so they could experience the outside world and be able to understand more complex messages from God. Materialists ridicule the idea of interiority in inanimate beings. They insist that nature is totally mindless without providing supporting, experimental data. However, Young's famous two-slit experiment, described below, suggests otherwise.

If we drop two pebbles a few inches apart into a quiet swimming pool, two wave circles appear, each having a top (crest) and a bottom (trough). At the point where the two crests meet, the wave is amplified. When crests and troughs from different waves meet, the waves are canceled out and the water is not disturbed. The same phenomenon occurs when light illuminates a screen with two slits, and it is also true with electrons, gamma rays, microwaves, and so on.

In an experiment with electrons, a barrier with two slits is placed in front of a phosphorescent screen such as that of a television. One electron at a time is shot at the screen. If both slits are open, the screen will show alternating series of fuzzy, dark, and bright diffraction bands (Figure 12-1). If only one slit is open, there will be one bright fuzzy band (Figure 12-2). If the experiment is performed with each slit closed for half the duration of the experiment, we see a double exposure separated by a distance that is equal to that between the slits (Figure 12-3). The patterns on the screen suggest that the electrons alter their travel patterns depending on the changing status of the slits. Based on this observation, we have to assume that the electrons somehow know whether the slit is open or closed.

Another step in the experiment is to determine which slit the electrons passed through. A particle detector is placed around one of the slits to record the path of the particles when the two slits are open and closed for half of the duration of the experiment (Figure 12-4). Although both slits are open, we do not see a series of dark and bright interference patterns as we have seen in Figure 12-1. Instead we see the pattern shown in Figure 12-3.

The second step of this experiment also suggests that the particles know that both slits are open. They are also able to sense the presence of the particle detector. The changes in their travel patterns confirm that inanimate beings have mind, as revealed in the Quran.

Moreover, no one can explain how an electron jumps from one orbit to another. Ernest Rutherford, a famous physicist, stated that "an electron would have to 'know' beforehand to which orbit it was going to jump. Otherwise, it would not emit light with a single, definite

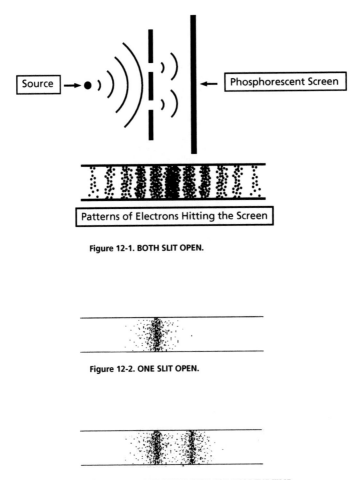

Figure 12-1. BOTH SLIT OPEN.

Figure 12-2. ONE SLIT OPEN.

Figure 12-3. EACH SLIT CLOSED FOR HALF THE TIME.

wavelength when it starts its leap."[4] Recently a few modern Christian scientists—for example, Alfred N. Whitehead and John F. Haught[5]—have accepted the presence of subjectivity and inner sense within all life forms, from single-celled organisms to the most complex human being. Whitehead accepts subjectivity for inanimate beings also.

God is a compassionate God who shows mercy for His creatures by voluntarily limiting His power so they can exercise free will to choose good over evil or vice versa. We have seen earlier in verse 10:99 that God has the power to make everyone a believer but did not. God commanded the Prophet (peace be upon them) not to coerce people

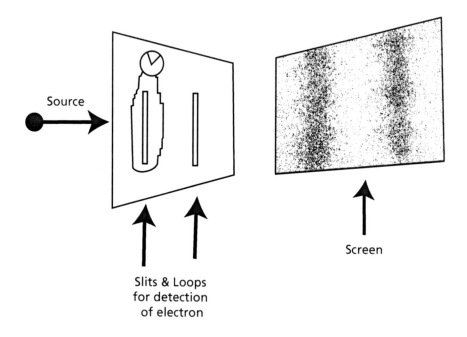

Figure 12-4. BOTH SLITS OPEN, WITH DETECTOR

to believe. The Quran also states: "…[know] that Allah is not a tyrant to His servants" (8:51) and "…Verily Allah never will change the condition of people unless they change it themselves (with their own soul)." (13:11) The Quran makes it clear that God is a guide, not a tyrant, and that He gives the freedom to His creatures to shape themselves and their future. We have witnessed His compassion when he did not compel the heavens, the earth, and the mountains to accept His offer. We acknowledge small and large DNA-based creatures as parts of a greater whole called life.

Similarly, inanimate objects such as stars and galaxies, the earth, the continents and subcontinents, states and nations, and so on are atomic aggregates. The collection of elements, such as fires, thunder, tornado, and floods as well as animate creatures—from genes to viruses, amoebas to humankind—are also atomic aggregates. Environments as tiny as a small culture plate (petri dish) with a growth of bacteria in a laboratory, or as huge as North America or

Asia, are parts of a whole called nature. The "Gaia hypothesis" of James Lovelock and Lynn Margulis maintains that the biosphere is in some sense a single "organism," a self-regulating whole that operates by its own laws.[6] I believe that it operates by Allah's laws. Based on the Quran, we must expand the definition of the biosphere to include the universe as a whole, under whose umbrella all individual units and subunits, or all separately identifiable units, are seen as organisms with subjectivities and selves.

Creatures can act within certain boundaries that God has specifically ordained for each subunit. For example, the electropositive ions potassium (K+) and sodium (Na+) cannot bond together to make a compound, while potassium or sodium are free to bond together with an electronegative ion, chloride (Cl-). God also created varying levels of inner sense and subjectivities with faculties to experience and respond, within whole systems and within their parts. Therefore, when He addresses His creatures, He mercifully expects different levels of responses from them.

We have seen earlier that inanimate units such as fire, thunder, etc. have duties and functions. Even "seventh heaven" has certain duties as stated in the verse, "... and He assigned to each heaven its duties and command...." (41:12) When nature or any of its parts responds to possibilities (proposals) in the arriving moments of the future from God, we interpret its actions in terms of natural selection. We observe nature deselecting the unfit choices of its own parts. In chapter 6, we saw Ibn Khaldun and al-Biruni attributing functions and duties to nature.

The complexity of information or future possibilities arriving for the use of any hierarchical order of being depends upon that order's inherent level of intellect and capability to actualize the information. Obviously, God did not want an amoeba to fly an airplane, but He wants human beings and *Jinn* to explore the heavens as clearly revealed in the following verse:

> **55:33** O company of *jinn* and men, were you able to penetrate bounds of heavens and the earth, do then penetrate (them). Yet, you shall not pass through save by an authority.

Creation/evolution of life is the result of God presenting possibilities (proposals) in each arriving moment of the future to the atoms as well as the aggregates of atoms. Within boundaries set by God, creatures, animate and inanimate, are free to actualize these possibilities into visible monuments of divine creation. In the Islamic universe, all possibilities—pleasant or unpleasant—are from God. The Quran states, "No misfortune befalls but by God's leave." (64:11) However, the actualization of good and bad from the arriving future belongs to His creatures.

The free choices of the whole and its subunits make our universe. In this unpredictable universe with an open future, creatures can get hurt but they can prosper, too. The pain and suffering of the components of the material universe is the price we pay for our free will. It is a place where we can experience life but where we will also be tested for the veracity of our faith in God through adversities, as the verse states:

> **2:155–157** And We shall test you with something of fear and hunger and loss of wealth, and lives and crops. Yet give good tidings to the patient.

We have to acknowledge that evolution is a fact. The evolution or unfolding of the universe is the process of creation by God. It did not come into being solely through coincidental causes in the past that created the future, as materialists claim. The physical universe and life evolved due to enterprising creatures armed with past experiences, choosing one or more possibilities within each moment flowing from the future intended by God. The lazy creatures who did not respond to the choices arriving from God through the messenger moments of the future remained what they were. Therefore, I feel no shame in accepting what I am as a result of my humanlike ancestors and the pre-human tailed creatures who preceded them and accepted Allah's offer to make them humans when He told the angels, "...I am setting on earth a vicegerent...." (2:30). In fact, I am proud of my pre-human ancestors' genes which chose to receive and grasp divine guidance in order to help our ancestors transform into humankind. Of course, the evolution of life into various species occurred as a result of His mercy to voluntarily limit His omnipotence and omniscience. Therefore, evolution is a fact of life as a result of God's gift of freedom to His creatures. Ian G. Barbour, physicist and Christian theologian, describes the process of creation/evolution splendidly in his book, *Religion and Science*. He compares God to a Choreographer of a dance in which the dancers are free to make many decisions by themselves.[7] In the creation/evolution of the Islamic universe, we dancers and musicians do not invent the better/worse or expanded/abridged versions of the dance or the symphony. By our own choice, we only objectify in our time and space any one of the variations of dance or symphony contained in the messenger-moments of the arriving future from God. This analogy of God as choreographer or composer also clarifies the Islamic doctrines that goodness and evil are from God and that humans actualize it by their choice.

Within the Islamic metaphysics of the future, the Islamic universe is always under God's providence and never leaves His hand, because He has the absolute power and free will to decide what possibilities to offer us. Humans as well as all other separately identifiable subunits of the universe have total freedom, based upon their individual competency, to choose and actualize the offered possibilities into visible monuments of Allah's creation. Just as for humankind, God has given a variety of freedoms to all subunits as well as the whole universe. Similar to the way of humanity, every subunit accepts or rejects God's guidance to choose from possibilities presented to them. All units freely select evil or righteous possibilities to get some benefit in this world or in the

Hereafter Universe (*al-Akhiram*). In such an unpredictable universe, some get hurt while others do not. (Please refer to chapter 6 for ample evidence from Jalaluddin Rumi to support this understanding.)

Based upon the teachings of the Quran, the far-reaching meaning of the phrase *Inshah Allah* (God so willing), and the concepts of Islamic metaphysics of the future, we see that while evolutionary accidents or contingencies appear random, they come from God as novel suggestions. The pivotal moment called the "present" is the opening into the material world. When messenger moments with novel possibilities or potentialities arrive at this door and a creature then objectifies them, we may see a mutation or the birth of a new species. An event in nature may appear random to humans because we have no knowledge of what the messenger-moment of the future has brought to every creature in the universe, and also because we also have no means to know what every individual being (i.e. separately identifiable components of the universe) has chosen to actualize into monuments of divine creation. Globally, the pain, suffering, or bliss that creatures experience are the normal outcome of the free choices creatures make among the possibilities they encounter at their individual levels when the future bursts into the razor-sharp edge of a fleeting present moment.

By granting freedom to His creatures to choose from the possibilities contained in the arriving messenger-moments of the future, God distanced Himself from qualities of a tyrant and distinguished Himself as the All Merciful (ar-Rahman) and The Evolver (al-Baari). He only proposes and creatures dispose. Finally, God is the source of all possibilities contained in the arriving future. Therefore, He becomes the Originator (al-Baadi), the Shaper (al-Musawwir), and the Evolver (al-Baari) of all things. The final result is that we live in a dynamic universe that is alive and in the process of creation in every moment, in the dazzling dance between Allah's limited creatures and His unlimited imagination.

Unlike Muslim Scientists, Gods Speaks Same Truth in and out of Mosques

WHAT I HAVE DISCUSSED so far is a systematic synthesis of science and Islamic theology. Most of it is supported by literal reading of the Quran and pre-Darwin Muslim scholars. However, contemporary traditional Muslim scholars reject the theory of evolution and many other theories of science. It is understandable. They have very limited education in science. At the same time, the rejection of the theory by Muslim scientists is disturbing.

Science has been defined as, "a continuing effort to discover and increase human knowledge and understanding through disciplined research. Using controlled methods, scientists collect observable evidence of natural or social phenomena, record measurable data relating to the observations, and analyze this information to construct theoretical explanations of how things work."

Based upon this definition, physicians, engineers, chemists, pharmacists, sociologists, and all other professionals who utilize the scientific method are called scientists. An embarrassing reality is that many Muslim scientists lead a double life. In their professional life they are true believer in science, but, when it comes to their religious lives, they passionately reject many scientific theories such as the theory of evolution and its applications. They have a truth tell in the mosques and another for their external professional lives. The old Latin expressions, "*contra evidential credo* (I believe despite the evidence)" and *credo quia evidentia* (I believe because of evidence)," most accurately differentiate their dual lives.

Scientists know, what they learned during formal education is but a basis onto which new knowledge is added as they progress though their professional lives. Once, more advanced remedies and method-ologies are discovered, most physicians do not apply the same remedies learned during formal education. Similarly, genuine scientists always apply new solutions in their professional life. Yet, how do the Muslim scientists fare regarding new knowledge within the context of religion?

Many Muslim scientists, relentless in questioning the validity of scientific opinions, will become submissive in accepting old and worn out commentaries, *tafsir,* of the Quran. Such intellectual complacency and lethargy is demonstrated despite the fact that many of the commentaries written in the past, some in a very remote past, conflict with current scientific knowledge that Muslim scientists use in their daily professional life. It is an amusing contradiction that many of our Muslim scientists accept science at one level but reject it on another. Why don't they apply the same intellectual curiosity to religion that they eagerly apply in other realms?

In order to reconcile this contradiction, many Muslim scientists cite the evolving nature of scientific theories. It is a fact that science has truth with proof but without certainty. It must be noted, though, that the science-rejecting Muslim scientist (It is an oxymoronic description) accepts religious truth with certainty but without proof. Many offer defensive rationalization that the scientific theories are not absolute, that the human intellect is not reliable, and the religious truth is absolute. In this justification, they forget that humans must utilize their intellect to determine whether the Hindu Geetha, or Jewish Torah or Christian Gospel, or Muslim Quran or all of them are to be acknowledged as holy books from God. If science-rejecting Muslim scientists have determined that human intellect cannot be considered a reliable measure of discernment, then why should anyone trust and accept their point of view in matters relating to religious and other human affairs including Muslim physicians' medical advice for their patients?

At the time of birth, babies have only instincts and reflexes. They show their need by crying and attracting those who feed them and comfort them. Mothers became the center of their universe. As babies get older their world-picture evolves, though still focused primarily on their mothers. Gradually this restricted world-picture matures and broadens to include family, friends, community, and the natural world. A world-picture is an echo of the underlining system of thought by which we process and harmonize our recurring experiences that we label as "facts."

Generally, a fact is defined as something that is true, something that actually exists, or something having objective reality that can be verified according to an established standard of evaluation. The more closely any particular fact is linked to the core of one's world-picture, the greater its subjective meaning and importance to us.

In other words, everything that is in harmony with our world-picture appears to us as truth. This interconnection between world-picture and truth can create profound positive and negative impact in any community, particularly in religious communities. The significances and outcome of the interconnection between world-picture and truth are illustrated in the history of Muslims. The arrival of religion of Islam transformed ordinary Arab life of 14th century into one of the most advanced civilization in less than two hundred years. George Sarton, a former Professor of History of Science at Harvard University, most elegantly describes this miracle of transformation

> "Briefest enumeration of the Arabic contributions to knowledge would be too long to be inserted here...The creation of a new civilization of international and encyclopaedic magnitude within less than two centuries is something that we describe, but cannot explain...Indeed the superiority of Muslim culture, say in

the eleventh century, was so great that we can understand their intellectual pride. It is easy to imagine their doctors speaking of western barbarians almost in the same spirit as ours do of the 'Orientals.' If there had been some ferocious eugenists among the Moslems they might have suggested some means breeding out all the western Christians and Greeks because of their hopeless backwardness. At that time Muslim pride would have been more conceivable because they almost reached their climax, and pride is never as great as when the fall is near. On the contrary only a few Christians were then aware of their inferiority; that awareness did not come upon them until much later—by the middle of thirteenth century."[1]

Similarly, according to British historian Robert Briffault gives the answer: while Europe wallowed in ignorance and barbarism, Muslim cities constituted "centers of civilization and intellectual activity."[2] The story of Muslims that Sarton and other historians described happened when free speech and critical thinking were part of Islamic community values. It is at a time al-Biruni wrote in *Vestiges of the Past (Athar-ul-Baqiya)*: "We must clear our minds...from all causes that blind people to the truth—old custom, party, spirit, personal rivalry or passion, the desire for influence."[3] It was the time when Muslims and the religion of Islam celebrated and practiced science and scientific method proudly. That was the world-picture and the truth then.

On the other hand, current Muslims have a different the world-picture and the truth. They reject science and scientific method in mosques. Our traditional scholars parrot uncritically what they heard from the schools with almost no academic freedom for dissensions. So, the "truth" reverberating all across Muslim world from the Muslim scholars and Imams whose world-picture is stuck in the past and who have limited modern knowledge of natural world, distort the meaning of a Quran when the Quran is meant to be timeless and ever-relevant. What was the end-result?

According to George Sarton, Christians in the Middle Ages were not even aware of their state of inferiority over Muslims ("On the contrary only a few Christians were then aware of their inferiority;

that awareness did not come upon them until much later—by the middle of thirteenth century."). Unfortunately, the role has reversed in the twenty first century. Muslims at large are unaware of their inferiority in scientific field and in other fields of knowledge.

How then do science-oriented Muslims best respond to science-rejecting Muslims who rally on the points that since science is always evolving, confirmed scientific observations could be proved wrong in the future? Should one reject the Quran if further scientific discoveries replace the present paradigm with a new one? Definitely not! If the Quran is the divine book in human vernacular, "the universe is a "written scroll"[4] (21:104) in the material medium or vernacular of the natural world.

The Quranic verses are called by Allah, *ayat*. The phenomena of the material world, Allah similarly refers to as *ayat*:

> **51:20–21** And in the earth are signs [*ayat*] for those who have firm faith, and in your own selves. Do you not discern?[5]

These two books, one the Quran and the other the book of nature come from the same divine source and so one divine revelation cannot contradict the other. According to the Quran, God provides humans with ample evidence in the universe to serve as proof of existence. If God told humans one thing in the Quran and gave the evidence for another in His book of Universe, this would do the opposite of proving His existence. Unlike Muslim Scientists, God Speaks Same Truth in and out of Mosques. So, an occasional contradiction has nothing to do with these two divine books but all to do with the meaning extracted from the two books. As this meaning can be distorted by our subjective world-view, it is vitally important that we seek always to find the most objective meaning that is relevant to all times, inclusive of new discoveries within the modern era.

As such when we encounter contradictions between the two divine books, our meaning and our world-picture are at fault for the imaginary conflict we have created between the two. Muslims must go back and re-read the two books in order to reconcile the contradictory meanings and their world-picture arising from their previous reading of the books. Such repeated experimentation, until the resolution of the conflict, is demanded by the Quran:

> **67:3–4** He created seven heavens in harmony. You do not see any disorder in the creation by the Gracious. Keep looking; do you see any flaw? Then Look again twice. Your eyes will come back humiliated and tired.[6]

Therefore, the Quran and the science applied to the natural world demand repeated experimentation and evidence so that they may come nearer to the truth until there is no conflict between the two divine books.

Finally, we, Muslims, talk today about the ʿQuranic truth' as the permanent kind of truth, the kind we swear by, the kind we feel in our bones to be true, the last truth of all that we are not willing to give up. It is that kind of truth, a universal belief shared by all Muslims, which has to be unraveled and restated harmoniously with each new discovery of verses written within the divine book of the universe in order for this truth to remain true for all the times. We shall not be afraid to tell the truth as it appears to us. Indeed, if we truly believe in God, we should recognize that Allah gave us the brains to conduct scientific research. Not to do so—not to use the marvelous gift of intelligence with which He has blessed us—would not only be impolite, it would express a heinous ingratitude to God. The Quran cautions the consequence of the squandering of divine gift offered to humanity in verse:

> **3:190–191** Behold! In the creation of heavens and the earth, and alternations of night and day,—There are indeed Signs for men of understanding,—Men who celebrate the praise of God, standing, sitting, and lying down on their sides, and contemplating the (wonders of) creation in the heavens and the earth: 'Our Lord! Not for naught hast Thou created (all) this! Glory to Thee! Give us salvation from the penalty of fire.'[7]

Although many contemporary Muslims believe that God instantaneously created a well-designed world, science has shown that the universe and life on earth evolved over billions of years. The Muslims in their classical period believed that creation is a process that occurred over long period of earthly time. According to the Quran and medieval Muslims, all separately identifiable small or large components of the universe have self and subjectivity. All inanimate material beings such as wind, fire, mountains etc., are living creatures. Our evolving universe is the result of the struggle between the small and the large components of the universe in which creatures fight each other to get some benefit and avoid some injury.

As such, the material world is not a perfect place of joy and jubilation. Our material world is characterized by the coexistence of evil and good; pain and joy. God intended it to be a fierce war zone, a testing ground for the evolution of His moral agents and a prelude to their entry to the Garden of Final Return. The Quran describes that, contrary to the imperfection of the transient material universe, the Hereafter world is a place devoid of evil, in which believers enjoy perfect peace: "And surely the world to come is greater in ranks, greater in preferment" (17:21)[8], and "Therein they shall hear no idle talk, no cause of sin, only the saying 'Peace, Peace!'"[9] (56:25)

Application of John Haught's novel and liberating concepts of "metaphysics of future" within the Islamic context unravels a logical solution to the conflict between science and contemporary Islam. It helps to understand the relationship between God's omniscience and human free-will also. I have explained that the expression *Inshah Allah* (God so willing) represents the Islamic metaphysics of the future. Messenger-moments of future do not exist and never arrive for material beings until messenger-moments emanate from God. They hold within them information that embodies potential possibilities or proposals that God grants to His creatures. These possibilities remain dormant until living beings exercise free will at a given moment to actualize divine proposals within a material medium. Thus, our evolving, short-lived, and imperfect material world is the result of the materialization of chosen possibilities coming from God by both small and large creatures alike. Yet, stunningly, in this potentially chaotic universe, which evolves through the practice of free choice, order emerges. Such order results from the interplay between the organization of possibilities that God makes within messenger-moments and the choices creatures make, based on their past experiences and self interest, before actualizing the possibilities visibly.[10]

I have argued that, according to the Quran, all separately identifiable beings have self and subjectivity. The creation is a process where messenger-moments, holding novel or routine potential for the future, arrive at the brief, living present of all creatures, allowing them to actualize the potential to reality. When potentials become visible realities of the material world as result of creatures' choices, the active present is kicked out to become "fixed past."[11] The proposals become visible monuments of divine creation. Deterministic scientists, however, believe that the "fixed past" created the universe. On the contrary, according to Islamic metaphysics the exercise of free will by material beings on the proposals arriving with the messenger-moments of future emanating from God create and evolve the universe. While God addresses the lower forms of being with the least complex and limited number of proposals, humans are presented with numerous, more complex possibilities. Each of us has the freedom to choose and objectify any of the proposals in the material of our world. Creation/Evolution is the end-result of the interplay between God reaching out "to the entire universe"[12] and the responses of material beings to the divine proposals.

The existence of evil or goodness within this world also comes from the choice made by the big and small creatures. Sometimes, choices made by creatures may totally block other creatures' freedom to make choices, leading to frustration for those who reject God. However, when all possibilities are blocked, resulting in calamity or even death, the believing creatures "seek help in patience and prayer," (2:153)[13] and say, "surely we belong to God, and to Him we return" (2:156).[14] In the defeat or submission during "the war in nature," true believers know that there exists "blessing and mercy from their Lord" for the patient among His creatures (2:157).[15]

The theory of evolution states that all life forms evolved from a common ancestor. Organisms that exist today are those that nature allowed to survive because of their adaptability to changing environmental circumstances. Materialists among the evolutionists contend that a random accident called mutation occurs in genes of some species, resulting in variations in species. Materialists argue that the contingent, accidental nature of these evolutionary variations is clear evidence against the Divine. At the same time the theory of evolution is no way anti-Islamic, but such an atheistic and antireligious interpretation of evolution has most probably led to contemporary Muslim clerics' (imams) and scholars' rejection of science. This rejection led in turn to a decline in scientific progress and subsequent underdevelopment in the Muslim world.

In mosques, imams preach that Adam and Eve were created in Paradise and expelled from it because the nuclear grandparents chose to trust grand Satan (*Iblis*) more than God. However, we have learned that the Quran and medieval Muslim scholars explained the concept of the evolution of life centuries before Charles Darwin. Even though we know today that the ape family and hominid family have a common ancestor historically, medieval Muslims were the first to entertain the idea of the human origin from the world of apes. Therefore, Adam could not have been created *ex nihilo* in Paradise, but raised from the earth as prophet.

God created the universe with classical laws of physics and chemistry to work at the macro-level of being, while quantum mechanics and indeterminacy rule at the micro—atomic and subatomic—levels. By such a design God maintains His omnipotence to respond to His creatures' needs. The generation of small quantum fluctuations in the micro-world can make major changes in the macro-world without violating the laws of classical physics and chemistry. Materialists, however, interpret God's actions at the quantum level as accidental genetic mutations.

The theology I have discussed blends the Quranic revelation, medieval Muslim understanding of nature, Whitehead's process thoughts, William Pollard's quantum physics theology, and John Haught's metaphysics of the future. Evidence for the story of the material universe supports an abrupt beginning at the Big Bang, then an evolution, then a final Big Crunch. God knows all possible numerous directions and roads that this universe and its component parts would take during their journey from Big Bang to Big Crunch, but He voluntarily and mercifully limited His omniscience and omnipotence in order to create freedom for His creatures to choose freely from possibilities that God presents to them with every approaching moment. The design of the universe with the indeterminate nature of quantum behavior of basic matter along with the blended theological interpretations explicated herein, allows us to conclude that God maintains His omnipotence, omniscience, love, and mercifulness without visibly violating any immutable laws of nature, allowing living beings to enjoy the gift of freedom. God speaks same truth in both of His books—the Quran and the universe. So, Muslim scientists must learn to speak same truth in and out of mosques as mandated by their faith.

Finally, how do the faithful respond to the scientism's argument that there is no empirical evidence for God?

All religions and science share certain common characteristics. Both have metaphysical commitments. Philosophers use the term "metaphysics" to refer to the general vision of reality that one holds to be true. Few metaphysical commitments of scientist are: 1. the five senses are the only portals of knowledge; 2. all wisdom comes from the material world; 3. scientific methodology is the only way to verify truth; and 4. the universe is intelligible. According to John F. Haught, a well-known Christian theologian, these metaphysical commitments of science are based only on the faith "because there is no way, without circular thinking, to set up a scientific experiment to demonstrate that every true proposition must be based in empirical evidence rather than faith."[16] For example, Einstein acknowledges that faith in an intelligible universe is essential to science itself. Therefore, Richard Dawkins, Daniel Dennett, and other followers of scientism initially have to produce empirical evidence for their own faith in the above scientific postulates without resorting to circular thinking before they ask from the faithful for the evidence for the existence God.

Finally, how do the faithful respond to the scientism's argument that there is no empirical evidence for God?

All religions and science share certain common characteristics. Both have metaphysical commitments. Philosophers use the term "metaphysics" to refer to the general vision of reality that one holds to be true. Few metaphysical commitments of scientist are: 1. the five senses are the only portals of knowledge; 2. all wisdom comes from the material world; 3. scientific methodology is the only way to verify truth; and 4. the universe is intelligible. According to John F. Haught, a well Christian theologian, these metaphysical commitments of science are based only on the faith "because there is no way, without circular thinking, to set up a scientific experiment to demonstrate that every true proposition must be based in empirical evidence rather than faith." (God and the New Atheism by John F. Haught). For example, Einstein acknowledges that faith in an intelligible universe is essential to science itself. Therefore, before Richard Dawkins, Daniel Dennett, and other followers of scientism ask for the evidence for the existence God, they have to produce empirical evidence for their own faith in the above scientific postulates.

Finally, I, being Muslim, is thankful to Stephen Hawkins for enlightening me with the divine method of creation of current and the Hereafter universe even though he believes that "It is not necessary to invoke God to light the blue touch paper and set the universe going;"[17] and to Charles Darwin for collecting the evidence to explain Allah's method of creation of human species. In order to understand the origin of the universe, both of them unknowingly carried out the following divine commandments that contemporary Muslims generally ignore:

> **29:20** "Say [O Muhammad]: 'Travel in the earth, then see, how Allah originated creation; and then He produces a later creation.' Allah is omnipotent."

ENDNOTES

NOTES TO INTRODUCTION

1. Moyers, Bill. *A World of Ideas,* p. 272.
2. Yahya, Harun. *The Evolution Deceit.*
3. Nasr, Seyyed Hussein. *Islamic Spirituality Foundation,* p. xxi.
4. Ibid. p. xxii.

NOTES TO CHAPTER 1

1. Barbour, Ian G. *Religion and Science,* p. 81.
2. Richard Dawkins. *The Blind Watchmaker.*
3. Richard Dawkins. *River out of Eden.* New York: Basic Books, 1995.
4. From Richard Lewontin's review of Carl Sagan's book, The Demon-Haunted World: Science as a Cradle in the Dark, in the New York Review of Books, January 9, 1997.
5. John F. Haught. *God After Darwin.* Boulder: Westview Press, 1999, p.86.
6. Ibid.
7. John F. Haught. *God After Darwin,* p. 83–88.
8. Guillanme, A. *The Life of Muhammad,* p. 375.
9. Barbour, Ian G. *Religion and Science,* p. 314

NOTES FOR CHAPTER 2

1. Nasr, Seyyed Hossein. *"Islam and the Environmental Crisis,"* MAAS Journal of Islamic Science. p. 35.

2. Eaton, C. G. *Islam and the Destiny of Man,* p. 87.
3. -Ghazzali, Imam Abu-Hamid. (1058-1119 a.d.). Ihya Ulum-Id-Din, p. 20.
4. Al-Suhrawardy, Allama Sir Abdullah Al-Mamun. The Sayings of Muhammad, p. 108.
5. Ibid. p. 114.
6. Ibn al-Hajjaj, Abul Husain Muslim (d.875). *Sahih Muslim,* trans. Abdul Hamid Siddiqui. Book 18, No.4261.
7. Bukhari, Muhammad ibn Ismail al-Jufi (d.897). *Sahih Al-Bukhari,* Vol. VI, p. 386.
8. Morris, Richard. *Time's Arrows,* p. 10.
9. *Encyclopaedia Britannica, Macropaedia,* "Time," Vol. 18, p. 411.
10. Morris, Richard. *Time's Arrows,* p. 66–67.
11. Hawking, Stephen W. *A Brief History of Time,* p. 7.
12. Morris, Richard. *Time's Arrows,* p. 27.
13. Ibid. p. 209.
14. Davies, Paul. God and the New Physics, p. 120.
15. March, Robert H. Physics for Poets, p. 124.
16. Sagan, Carl. Cosmic Connections, p. 246.
17. March, Robert H. Physics for Poets, p. 125.
18. Morris, Richard. Time's Arrows, p. 178.
19. Ibid. p.180–181.
20. Asad, Muhammad. The Message of the Qur'an, p. 513.
21. Schroeder, Gerald L. The Science of God, p. 164.
22. Ibid. p. 164
23. Ibid. p. 164.
24. Davies, Paul. God and the New Physics, p.128.
25. Hakim, Abdul Hakim. The Metaphysics of Rumi, p 18.
26. Whinfield, E. H. Teachings of Rumi, Inside cover page.

NOTES TO CHAPTER 3

1. Hardy, Edward R. "Origin and Evolution of Universe." Macropaedia, Vol. 18, p 1007.
2. Adler, Mortimer J. , p. 187.
3. Eve, Raymond A., and Harrold B. Francis. p. 13–14
4. Ibid. p.48.
5. Ibid. p. 48.
6. Ibid. p. 47–48.
7. Ibid. p. 46.
8. Dods, D.D, Marcus, and others. "Book of Genesis" in , Vol. 1. p. 5–7.
9. Isaac. Beginnings: , p. 253.

10. Hardy, Edward R. "Origin and Evolution of Universe." Encyclopaedia Britannica, Macropaedia, Vol. 18, p 1007

11. Abell, George O. "The Age of the Earth and the Universe" in Godfrey, Laurie, ed., p. 46

12. Sagan, Carl. , p. 246.

13. Ferris, Timothy (narrator). 1990. PBS Home Video, Beverly Hills.

14. Hawking, Stephen W. p. 70-72.

15. Barrow, John D., and Silk, Joseph. 1983. p. 73–75

16. Hawking, Stephen W. p. 70–72

17. Ferris, Timothy (narrator). 1990. PBS Home Video, Beverly Hills.

18. Asimov, Isaac. Beginnings: The Story of the Origins—Of Mankind, Life, the Earth, the Universe, p. 263–264.

19. Sagan, Carl. p. 246

20. Asad, Muhammand, p. 491.

21. Lane, Edward William. , part 2, p. 816.

22. Sagan, Carl. p. 246

23. Davies, Paul. p. 204.

24. Jastrow, Robert. p. 129.

25. Schroeder, Gerald L. 1990. p. 81

26. New York: Watch Tower Bible & Tract Society of Pennsylvania. p. 186.

27. Hardy, Edward R. "Origin and Evolution of Universe." Macropaedia, Vol. 18, p. 1011.

28. Davies, Paul. p. 204

29. Davies, Paul. p. 209

30. Sagan, Carl. p. 231

31. Ibid.

32. Lane, Edward. part 4, p. 1308.

33. Sagan, Carl. p. 231

34. Davies, Paul. p. 204

35. Salahi, Adil M. and Ashur A. Shamis. 1979.

36. Hardy, Edward R. "Origin and Evolution of Universe." Macropaedia, Vol. 18, p. 1011.

37. Kaufman III, William J. p. 84.

38. Hawking, Stephen W. p 133.

39. Hardy, Edward R. "Origin and Evolution of Universe." Macropaedia, Vol. 18, p. 1011.

40. Sagan, Carl. p. 267.

NOTES TO CHAPTER 4

1. Edey, Maitland A. and Donald C. Johanson. Blue Print, pp. 34, 291.

2. Al-Faruqi, I. R. The Cultural Atlas of Islam, p. 313.

3. Bronowski, Jacob. p. 166.

4. Briffault, Robert. 1980. pp. 200–201.
5. al-Biruni, Abu Raihan. 1000 A.D. Edward C. Sachu, trans., p. 85.
6. Schroeder, Gerald L. p. 31
7. Asimov, Isaac. pp. 22–23.
8. Edey, Maitland A. and Donald C. Johanson. p. 8.
9. Eve, Raymond A. and Francis B. Harrold. p. 46.
10. Asimov, Isaac. Beginnings: The Story of the Origins—Of Mankind, Life, the Earth, the Universe, p. 26.
11. al-Biruni, Abu Raihan. 1000 A.D. Edward C. Sachu, trans., p. 16–22.
12. Newport, Frank, "One-Third of Americans Believe the Bible is Literally True", 25 May 2007, Gallup, May 31 2010, http://www.gallup.com/poll/27682/onethird-americans-believe-bible-literally-true.aspx
13. Eve, Raymond A. and Francis B. Harrold. p. 4.
14. Trefil, James. Reading of the Mind of God, p.135.
15. Abell, George O. "The Age of the Earth and the Universe," in Godfrey, Laurie, ed., p. 35
16. Ibid. p. 43.
17. Marvin, Ursula B. "Continental Drift," Macropaedia, (15th ed.) Vol. 5. p 108.
18. Abell, George O. "The Age of the Earth and the Universe," in Godfrey, Laurie, ed., p. 36.
19. Micropaedia, (15th ed.) "Al-Biruni," Vol. II. p. 42.
20. al-Biruni. Kitab Tahdid al-Amakin Listashiah Masafat al-Masakin (The Determination of the Coordinates of Positions for Correction of Distances between Cities). Jamil Ali, trans., pp.14–16.
21. al-Biruni, Abu Raihan. 1000 AD. Edward C. Sachu, trans., p. 30
22. Ali, Ameer. p. 431–432.
23. Plott, John C. pp. 258–259.
24. Nasr, Sayyed Hossein. p. 80.
25. Micropaedia, "Avicenna," Vol. I. p.661.
26. Osborn,H. F. , p. 76.
27. Dodson, Edward O. and Peter Dodson. p. 79.
28. Shadewald, Robert J. "The Evolution of Bible-science," in Godfrey, Laurie, ed., 286
29. Gould, Stephen Jay. pp. 119–125.
30. al-Biruni. Edward C Sachau, trans. Vol. I, p. 198.
31. Nasr, Sayyed Hossein. p. 141.
32. al-Biruni. Edward C. Sachau, trans. Vol. I, pp. 378–379.
33. al-Biruni, Abu Raihan. 1000 a.d.. Edward C. Sachu, trans., p. 27–28.
34. Nasr, Sayyed Hossein. , pp. 80-81.
35. Ibid. p. 245.
36. Ibn-Khaldun, Abdul-Rahman. Franz Rosenthal, trans. Abr.ed.
37. Ibid.

38. Ibn-Khaldun, Abdul-Rahman. . Franz Rosenthal, trans. Vol. I, pp. 169–170.
39. Ibid. p. 173
40. Edey, Maitland A and Donald C. Johanson. p. 8.
41. Asimov, Isaac. Beginnings: The Story of the Origins—Of Mankind, Life, the Earth, the Universe, pp. 46–48.
42. Nasr, Sayyed Hossein. p. 142.
43. Osborn, H. F. p. 76.
44. Draper, John William. p. 42.
45. Briffault, Robert. p. 188–190.
46. Ibid 190.

NOTES TO CHAPTER 5

1. Ashley, Montagu (ed.), "Evolution as Fact and Theory," by Stephen Jay Gould, pp. 118–119.
2. Curtis, Helena and Sue N. Barnes. (5th ed), p. 974
3. Shapiro, Robert. "Probing of the Origin of Life," p. 9
4. Ibid. p.14.
5. Miller, Stanley L., and Robert G. Orgel. p. 33.
6. Shapiro, Robert. "Probing of the Origin of Life," p. 14–15.
7. Abelson, Philip H. "Chemical Events on the Primitive Earth," Vol.55, p.1365.
8. Dimroth, Erich, and Michael M. Kimberley. "Pre-Cambrian atmospheric oxygen: Evidence in the sedimentary distributions of Carbon, Sulfur, Uranium, and Iron," Vol. 13, September 1976, p. 1161.
9. Henderson-Sellers, A. Benlow and A. J. Meadows. "The Early Atmospheres of Terrestrial Planets," Vol. 21 (1980), p. 81.
10. Shapiro, Robert. p. 112.
11. Doolittle, Russell F. "," Scientists Confront Creationism, Godfrey, Laurie R., ed., p. 89–90.
12. Shapiro, Robert. "Probing of the Origin of Life," p. 22.
13. Sagan, Carl E. "Life," 15th ed., p. 902.
14. Ferris, James. P and Kamaluddin. Oligomerization Reactions of Montmorinollite Clay: The Effect of Mononucleotide Structure on Phosphodiester Bond Formation. Journal of Origins of Life and Evolution of Biospheres. Volume 19, No. 6. November, 1989.
15. Shapiro, Robert. p. 164.
16. Edey, Maitland A. and Donald C. Johanson. p. 288–289
17. Darwin, Charles. pp. 73–74.
18. Ibid., p. 450
19. Dodson, Edward O. and Peter Dodson. p. 189-190.
20. Ibid., p 233.

21. Gould, Stephen Jay pp. 396–397.

22. Shermer, Michael. "25 Creationists' Arguments & 25 Evolutionists' Answers," Vol.2, No.2.

23. Futuyma, Douglas J. "World Without Design." March 1987, p. 34.

24. Dodson, Edward O. and Peter Dodson. p. 543.

25. Mayr, Ernst, and William Provine, eds. "Prologue: Some Thought on the History of the Evolutionary Synthesis," p. 1.

26. Darwin, Charles. p. 312.

27. Stanley, Steven M. p. 39.

28. Raup, David M. "Conflict between Darwin and Paleontology," Field Museum of p.25.

29. Gould, Stephen Jay. "Evolution, As Fact And Theory," Montagu, Ashley, ed., p.123.

30. Gould, Stephen Jay. "A quahog is a quahog," August-September 1979, p. 18–26.

31. Futuyma Douglas J. "The Theory of Evolution," (15th ed.), p. 1005.

32. Dodson, Edward O. and Peter Dodson. Evolution Process and Products, p. 178.

33. Gould, Stephen Jay and Nils Eldredge. "Puctuated Equilibrium Comes of Age," Nature (1993), p. 223.

34. Futuyma, Douglas J. "The Theory of Evolution," Encyclopaedia Britannica (15th ed.), p. 986.

35. Ibid. p.987

36. Ibid.

37. Dodson, Edward O. and Peter Dodson. Evolution Process and Products, p. 52–53.

NOTES TO CHAPTER 6

1. Blackmore, Vernon., Page, Andrew. Evolution the Great Debate, Illinois: Lion Publishing Company, 1989, p. 106.

2. Ibid. 102

3. Gould, Stephen Jay. "Knight Takes Bishop?" Natural History, (May 1986), p. 18.

4. Lucas, J. R. Historical Journal, XXII (1979), p. 102

5. Draper, John William. The Conflict between Religion and Science, pp. 187–188.

6. Briffault, Dr.Robert. The Making of Mankind, p. 201.

7. Godfrey, Laurie R (ed.). Scientists Confront Creationism, "Scopes and Beyond: Antievolution and American Culture," Cole, John R, p. 25–27.

8. Draper, John William. The Conflict between Religion and Science. p. 118.

9. Mayr, Ernst. One Long Argument: Charles Darwin and the Genesis of Modern Evolutionary Thought, p. 3.
10. Miller, Richard. The Encyclopedia of Evolution, p. 116.
11. Hitti, Philip K. History of Arabs, p. 588.
12. Draper, William John. The Intellectual Development of Europe, Vol. 2, pp. 39–40.
13. Durant, Will. The Story of Civilization, Vol. 4, p. 246-249.
14. Tufail, Abu Bakr Muhammad. Hai bin Yaqzan (The Journey of The Soul), pp. vi–vii.
15. http://www.english.upenn.edu/Projects/knarf/Darwin/temple1.html
16. Durant, The Story of Civilization, Vol. 4, p. 243.
17. Ibid. 249.
18. Khaldun, Ibn. The Muqaddimah, trans. by Franz Rosenthal, Vol. 1, p. 195.
19. Ibid. Vol. 1, p. 173.
20. Ibid. Vol. 3. p. 238
21. Ibid. Vol. 2, p. 422–423.
22. Hussani, S.A.Q. The Pantheistic Monism of Ibn Arabi, p. 61.
23. Ibid. p. 66.
24. Ibid p. 107.
25. Ibid. p. 106.
26. Ibid. p. 61
27. Ibid. p.105
28. Ibid. p. 105.
29. Gould, Stephen Jay. Ever Since Darwin, p. 60–61.
30. Biruni, Al (1000 A.D) The Athar-Ul-Bakiya (The Chronology of Ancient Nations), translated by Dr. C .Edward Sachu, p.98.
31. Ikhwan al-Safa. Rasa'il Ikhwan al-Safa wa-Khull an al-Wafa (Epistles of Brothrens of Purity), Vol. II, p. 178.
32. Ibid. Vol. II, p. 181.
33. Ibid. Vol. II, p. 182.
34. Ibid. Vol. II, p. 192.
35. Nicholson, Reynold A. Rumi, p. 11.
36. Hakim, Dr Khalifa Abdul. The Metaphysics of Rumi, p. 36.
37. Ibid. p. 36
38. Ibid. p. 37.
39. Al-Biruni. Kitab al-Jamahir fi ma'rifat al-jawahir, (Arabic), Hyderabad: Dairatu'l Ma'arif'l Osmania, 1934, p. 80.
40. Tufail, Abu Bakr Muhammad. Hai bin Yaqzan (The Journey of The Soul), pp. vi–vii.
41. Osborn.H.F. From Greeks to Darwin, pp. 77–78.
42. Shadewald, Robert. National Center For ScienceEducation Reports, Vol. 10 (March–April 1990), p. 21.

43. Encyclopaedia Britannica, (15th ed.) Micropaedia. "Ibn Miskawayah." Vol. V, p. 272.

44. Iqbal, Sir Mohammad. The Reconstruction of Religious Thought in Islam, p. 133-134.

45. Durant, The Story of Civilization, Vol. 4, p. 238.

46. Ibid. p. 239.

47. Durant, p. 245.

48. Ibid. p. 289.

49. Ibid. p. 288.

50. Ali, Ameer. The Spirit Islam, p. 424

51. Draper, John, The Intellectual Development of Europe, Vol. II, p.48.

52. Encyclopaedia Britannica, (15th ed.), Micropaedia, "Nazzam, Ibrahim," Vol. VII. p. 233.

53. Hakim, Dr Khalifa Abdul. The Metaphysics of Rumi, p. 34.

54. Khaldun, Ibn. The Muqaddimah, trans. by Franz Rosenthal, Vol. 1, p. 173.

55. Gould, Stephen Jay. Ever Since Darwin. p. 44–45.

56. Darwin, Charles. The Origin of Species. p. 49, 74.

57. Biruni, Al. (1000 a.d.) Fi Tahqiq Ma Li'l-Hind (Alberuni's India), Translated by Dr.Edward C. Sachu, p. 400.

58. Ibid. p. 400.

59. Darwin, Charles. The Origin of Species. p. 75, 86, 88

60. Biruni, Al. The Athar-Ul-Bakiya (The Chronology ofAncient Nations), Translated by Dr.C. Edward Sachu, p. 92–93.

61. Ibid. p. 295.

62. Hakim, Dr Khalifa Abdul. The Metaphysics of Rumi, Ibid. p. 39.

63. Ibid. p. 37–38.

64. Gould, Stephen Jay. Wonderful Life. New York: W.W. Norton & Company, Inc. 1989, p. 320.

65. Stanfield, William D. The Science of Evolution, New York: The Macmillan Company, 1977. p. 26.

66. Al-Suhrawaardy, Allama Sir Abdullah. The Sayings of Muhammad, p. 94.

NOTES TO CHAPTER 7

1. Al-Halveti, Sheikh Tosum Bay Bayrack Al-Jerrahi. The Most Beautiful Names, p. vii–viii.

2. Azad, Mawlana Abdul Kalam. Tarjuman al-Qur'an, Trans. Syed Abdul Latif, Vol. I. p. 19.

3. Ali, Maulana Muhammad. The Religion of Islam, p. 135.

4. Lane, Edward William. Arabic-English Lexicon, part 1. p. 7

5. Ibid. Part 2, p. 799

6. Ibid. Part 2, p. 799
7. Ibid. Part 2, p. 800
8. Ibid. Part 2, p. 799–800
9. Ali, Yusuf. The Qur'an. p.1529
10. Ibid.
11. Ibid.
12. Lane, Edward William. An Arabic-English Lexicon, part 4, p. 1745.
13. Ali, Yusuf. The Qur'an, Verse 10–4, p. 484
14. Bucaille, Maurice. What is the Origin of Man, p. 173.
15. Lane, Edward William. Arabic-English Lexicon, part 6, p. 2350.
16. Gould, Stephen Jay. Ever Since Darwin, p. 61.
17. Lane, Edward William. Arabic-English Lexicon. part 3, p. 1142.
18. Hakim, Khalifa Abdul. The Metaphysics of Rumi, p.36.
19. Lane, Edward William. An Arabic-English Lexicon, part 4, p. 1397.
20. Lane, Edward William. An Arabic-English Lexicon. part 3. p. 842.
21. Khaldun, Ibn. The Muqaddimah. Vol. 2. p. 424.
22. Lane, Edward William. An Arabic-English Lexicon, Part 5, p. 1973.
23. Lane, Edward William. Arabic-English Lexicon. Part 8. 2791.
24. Razi, Fakr al-Din (1149–1209 a.d.). Lebanon: Dar al-Fikar. Al Tafsir al-Kabir. Vol. 7. p. 212.
25. Lane, Edward William. Arabic-English Lexicon. Part 2. p.792.
26. Al-Biruni, Athar-ul-Bakiya (Chronology of Ancient Nations), p. 98
27. Al-Jahiz (a.d. 776–869). Chance or Creation, Translated by M. A. S. Abdel Haleem, pp. 113–114.
28. Ibn Khaldun. The Muquddimah, Trans. Franz Rosenthal, Vol. 1. p. 195.
29. Russell, Bertrand. Religion and Science, p. 80.
30. Sagan, Carl. Cosmic Connections, p. 246.
31. Al-Biruni. Kitab Tahdid Nihayat al-Amakin LitashiahMasafat al-Masakin (The Determination of the Coordinates of Positions for the Correction of Distances between Cities), Trans by Jamil Ali, p. 15-16.
32. Davies, Paul. God and the New Physics, p. 101
33. Ibid. p.101–102.
34. Ibn Khaldun. The Muqaddimah, Vol. 1, p. 173.
35. Milner, Richard. The Encyclopedia of Evolution.

NOTES TO CHAPTER 8

1. Bukhari, Muhammad ibn Ismail al-Jufi. (d. 897). Sahih Al-Bukhari. Translated by Muhammd Muhsin Khan. Chicago: Khazi Publications, 1977. Vol 8, Book 77, Number 593.

NOTES TO CHAPTER 9

1. Niazi, Maulana Kauser. Creation of Man, p. 88.
2. Asad, Muhammad. The Message of the Qur'an, p. 9.
3. Maududi, Abu A'la S. The Meaning of the Qur'an, Verse 32:17. Vol. 1, p. 51.
4. Maududi. The Meaning of the Qur'an, Verses 37: 7–10, Vol. 12, p.65.
5. Lane, Edward William. An Arabic-English Lexicon, Part 1, p. 37.
6. Asad, Muhammd. The Message of the Qur'an, p.205.

NOTES TO CHAPTER 10

1. Rippin, Andrew and Knappert, Jan. Textual Source for the Study of Islam, pp. 116–121.
2. Asad, Muhammad. The Message of the Qur'an, p.930.

NOTES TO CHAPTER 11

3. Gribbin, John. In Search of Schrodinger's Cat. p. 5.
4. Davis, Paul. God and the New Physics, pp. 101–102.
5. Russell, Robert John. "Special Providence and Genetic Mutation: A New Defense of Theistic Evolution," Evolutionary and Molecular Biology, p. 202.
6. Gribbin, John. In Search of Schrodinger's Cat. p. 61.
7. Miller, Kenneth R. Finding Darwin's God, p. 209.
8. Nasr, Seyyed Hossein. Islamic Spirituality Foundations, p. xxi.
9. Barbour, Ian G. Religion And Science, p.187.
10. Russell, Robert J. "Theistic Evolution: Does God really Act in Nature?" Center for Theology and the Natural Science Bulletin, 1995, 15.1.
11. Miller, Kenneth R. Finding Darwin's God, p. 207.

NOTES TO CHAPTER 12

1. Hakim, Dr Khalifa Abdul. The Metaphysics of Rumi, p. 35.
2. Ali. Sayed Anwar. Qur'an The Fundamental Law of Human Life, Vol. 11, pp. 362–364.
3. Ibid
4. Morris, Richard. The Nature of Reality, p. 15.
5. Haught, John F. God After Darwin, pp. 176–180.
6. Lovelock, James, E. Gaia, A New Look at the Earth.
7. Barbour, Ian G. Religion And Science, p. 314.

NOTES TO SUMMARY

1. George Sarton: "The History of Science and The New Humanism." Page 87–90.

2. Briffault, Robert. The Making of Mankind, p. 188–190.

3. Durant, The Story of Civilization, Vol. 4, p. 243.

4. Arberry. A. J. 1955. The Koran Interpreted.

5. Khatib, Dr. M. M. The Bounteous Koran.

6. Yuksel, Edip; al-Shaiban, Layth Saleh; Schulte_Nafeh. Quran A Reformist Translation.

7. Ali, Yusuf. The Holy Quran

8. Khatib, Dr. M. M. The Bounteous Koran 299. Arberry, Arthur J. The Koran Interpreted,

9. Arberry, Arthur J. The Koran Interpreted, 300. Barbour, Ian G. Religion and Science, p. 296

10. Barbour, Ian G. Religion and Science, p. 296.

11. Haught, John F. God After Darwin, p.103

12. Ibid. p. 99.

13. Khatib, Dr. M. M. The Bounteous Koran, Verse 2:153.

14. Ibid. Verse 2:156.

15. Ibid. Verse 2:157.

16. Haught, John F. God And The New Atheism A Critical Response to Dawkins, Harris, and Hittches.

17. Hawking, Stephen and Mliodinow, Leonard. The Grand Design.

BIBLIOGRAPHY

Abelson, Philip H. "Chemical Events on the Primitive Earth." *Proceedings of National Academy of Science*, Vol. 55 (1966), p. 1365-1372.

Adler, Mortimer J. *Aristotle for Everybody*. New York: Macmillan Publishing Co. Inc, 1978.

Al-Biruni. (d. 1048 AD.) *The Athar-Ul-Bakiya (The Chronology of Ancient Nations)*. Translated by Dr. C. Edward Sachu. London: W. H. Allen, 1879.

Fi Tahqiq Ma Li'l-Hind (Alberuni's India). Translated by Dr. Edward C. Sachu. London: Kegan Paul, Trench, Trubner & Co. Ltd., 1914.

Kitab al-Jamahir fi ma'rifat al-jawahir, (Arabic). Hyderabad: Dairatu'l Ma'arif'l Osmania, 1934.

Kitab Tahdid al-Amakin Listashiah Masafat al-Masakin (The Determination of the Coordinates of Positions for Correction of Distances between Cities). Translated by Jamil Ali. Beirut: The American University of Beirut, 1967. al-Faruqi, I. R. *The Cultural Atlas of Islam*. New York: Macmillan Publishing Company, 1986.

Al-Ghazzali, Imam Abu-Hamid. (1058-1111 AD.) *Ihya Ulum-Id- Din*. Translated by Al-Haj Maulana Fazal-Ul-Karim. Lahore. Pakistan: Book Lovers Bureau, 1971.

Ali, Ahmed. *Al-Qur'an*. Princeton: Princeton University Press, 1990.

Ali, Ameer. *The Spirit of Islam*. Lahore, Pakistan: Pakistan Publishing House, 1981.

Ali, Maulana Muhammad. *The Religion of Islam*. Lahore, Pakistan: Ahmadiyyah Anjuman Ish'at Islam, 1950.

Ali, Sayed Anwar. *Qur'an, The Fundamental Law of Human Life*. Karachi, Pakistan: Hamdard Foundation Press, 1987.

Ali, Yusuf. *The Holy Qur'an*. Indianapolis: American Trust Publications, 1977.

Al-Halveti, Sheikh Tosum Bay Bayrack Al-Jerrahi. *The Most Beautiful Names.* Vermont: Threshold Books, 1985.

Al-Jahiz. (AD 776-869.) *Chance or Creation.* Translated by M. A. S. Abdel Haleem. Berkshire, U.K: Garnet Publishing Limited, 1995.

Allman, William F. *"The Origin of Modern Humans.* WHO WE WERE." U.S. News & World Report. Vol. 111 (1991), No. 12.

Al-Suhrawaardy, Allama Sir Abdullah. *The Sayings of Muhammad.* New York: Carol Publishing Group, 1990.

Ansari, Muhammad Fazl-ur-Rahman. *The Qur'anicFoundations and Structure of Muslim Society.* Karachi, Pakistan: The World Foundation of Islamic Mission, 1977.

Arberry, Arthur J. *The Koran interpreted.* New York: Collier Books, 1955.

Asad, Muhammad. *The Message of the Qur'an.* Gibraltar: Dar Al- Andalus, 1980.

Ashley, Montagu, ed. *Science and Creationism.* Oxford, New York, Toronto, Melbourne: Oxford University Press, 1984.

Asimov, Isaac. *Beginning: The Story of Origins—of Mankind, Life, the Earth, the Universe.* New York: Walker & Co., 1987.

Azad, Mawlana Abdul Kalam. *Tarjuman al-Qur'an.* Translated by Syed Abdul Latif. Hyderabad, India: Dr. Syed Abdul Latif's Trust for Qur'anic and Other Cultural Studies, 1981.

Barbour, Ian G. *Religion and Science.* New York: HarperCollins, 1997.

Barrow, John D., and Joseph Silk. *The Left Hand of Creation.* New York: Basic Books Inc, 1983.

Blackmore, Vernon, and Andrew Page. *Evolution the Great Debate.* Illinois: Lion Publishing Company, 1989.

Briffault, Robert. *The Making of Mankind.* Lahore, Pakistan: Islamic Book Foundation, 1980.

Bronowski, Jacob. *The Ascent of Man.* London, U.K: The British Broadcasting Corporation, 1981.

Bucaille, Maurice. *What is the Origin of Man?* Paris, France: Seghers, 1983.

Bukhari, Muhammad ibn Ismail al-Jufi. (d. 897). *Sahih Al-Bukhari.* Translated by Muhammd Muhsin Khan. Chicago: Khazi Publications, 1977.

Cavalli-Sforza, Luigi Luca. "Genes, Peoples and Languages." *Scientific American.* 265.5 (1991): 104-106.

Curtis, Helena and Sue N. Barnes. Biology, (5th ed.). New York: Worth Publishers, 1989.

Daniel Dennett. *Darwin's Dangerous Idea: Evolution and the meaning of life.* New York: Simon & Schuster, 1995.

Darwin, Charles. *The Origin of Species.* New York, Scarborough: New American Library, 1958.

Davies, Paul. *God and the New Physics.* Simon and Schuster, Inc., 1983.

Dawood, N. J. *The Koran.* New York: Penguin Books, 1990.

Dimroth, Erich, and Michael M Kimberley. "Pre-Cambrian atmospheric oxygen: Evidence in the sedimentary distributions of Carbon, Sulfur, Uranium, and Iron." *Canadian Journal of Earth Sciences*, XIII (September 1976): 1161-1185.

Dods, D.D., Marcus, and others. "Book of Genesis" in *An Exposition of the Bible*. Hartford: The S. S. Scranton Co, Vol. 1. 1903.

Dodson, Edward O., and Peter Dodson. *Evolution: Process Product*. Boston: Prindle, Weber & Schmidt, 1985.

Draper, John William. *The Conflict between Religion and Science*. New York: D. Appleton and Company, 1875.

Draper, William John. *The Intellectual Development of Europe*. New York: London: Harper & Brother, 1876.

Durant, Will. *The Story of Civilization*. Vol. 4. New York: Simon and Schuster, Inc., 1950

Eaton, C. G. *Islam and the Destiny of Man*. New York: Macmillan, 1985.

Edey, Maitland A., and Donald C. Johanson. *Blue Print*. Boston, Toronto, London: Little, Brown and Company, 1989.

Elissi, Muhammad 'Abdul Haleem. *Holy Qur'an*. Hyderbad, India: The Barney Academy, 1981.

Erwin, Douglas H., and James W. Valentine. "Hopeful Monsters, Transpons, and Metazoan Radiation." *Proceedings of National Academy of Science*. Vol.81: 5482-5483, 1984.

Encyclopaedia Britannica-Micropaedia. 15th ed. 1981.

Encyclopaedia Britannica-Macropaedia. 15th ed. 1981.

Eve, Raymond A., and Francis Harrold. *The Creationist Movement in Modern America*. Boston: Twayne Publishers, 1991.

Ferris, Timothy (Narrator). "*The Creation of the Universe*." (PBS Home Video. Beverly Hills, 1990.

Futuyma, Douglas J. "World Without Design." *Natural History*. (March, 1987): 34-36.

"The Theory of Evolution." *Encyclopaedia Britannica*. 15th ed., pp 981-1009, 1988

Gobons, Ann. "Human Evolution: Y Chromosome Shows That Adam Was an African." *Science*. 278.10 (1997): 804-805.

Godfrey, Laurie R., ed. *Scientists Confront Creationism*. New York, London: W. W. Norton & Company, 1983.

Gould, Stephen Jay. *Time's Arrow Time's Cycle*. Cambridge, Massachusetts, London, U.K: Harvard University Press, 1987.

"A quahog is a quahog." *Natural History*, (August- September 1979): 18-26.

Gould, Stephen Jay, and Nile Eldredge. "Puctuated Equilibrium Comes off Age." *Nature* 336(1993): 223-227.

Gould, Stephen Jay. *Eight Little Piggies*. New York, London: W. W. Norton & Company, 1993.

Ever Since Darwin. New York, London: W. W. Norton, 1977.

"Knight Takes Bishop?" *Natural History*, (May 1986): 18-33.

Wonderful Life. New York: W. W. Norton & Company, 1989.

Guillanme, A. *The Life of Muhammad*. Oxford: Oxford University Press, 1955.

Hakim, Dr Khalifa Abdul. *The Metaphysics of Rumi*. Lahore, Pakistan: Institute of Islamic Culture, 1977.

Hardy, Edward R. "Origin and Evolution of Universe," *Encyclopaedia Britannica—Macropaedia*, 15th ed., Vol. 18: 1007-1011.

Haught, John F. *God After Darwin*. Boulder: Westview Press, 1999.

Hawking, Stephen W. *A Brief History of Time*. Toronto, London, New York, Sydney: Bantam Books, 1988.

Henderson-Sellers A., A. Benlow, and A. J. Meadows. "The Early Atmospheres of Terrestrial Planets." *Quarterly Journal of Royal Astronomical Society*. Vo.21 (1980): 74-81.

Hitti, Philip K. *History of Arabs*. New York: St. Martin's Press, 1970.

Hussani, S. A. Q. *The Pantheistic Monism of Ibn Arabi*. Lahore, Pakistan: SH. Muhammad Ashraf, 1979.

Ibn al-Hajjaj, Abul Husain Muslim (d. 875). *Sahih Muslim*. Book 18, No. 4261. Translated by Abdul Hamid Siddiqui. New Delhi: Kitab Bhavan, 2000. (Available on the Web: *www.usc.edu/dept/MSA/fundamentals/hadithsunnah/muslim/*)

Ibn-Khaldun, Abdul-Rahman (d. 1406). *Muqaddimah*. Translated by Franz Rosenthal. Princeton: Princeton University Press, 1980.

The Muqaddimah (Abridged edition). Translated by Franz Rosenthal. Princeton: Princeton University Press.

Ikhwan al-Safa. *Rasa'il Ikhwan al-Safa wa-Khull an al-Wafa* ("*Epistles of Brothrens of Purity*"), (Arabic). Beirut, Lebanon: Dar-ul-Beirut Publishers, 1983.

Iqbal, Sir Mohammad. *The Reconstruction of Religious Thought in Islam*. Lahore, Pakistan: Ashraf Printing Press, 1982.

Irving, T. B. *The Qur'an*. Brattleboro: Amana Books, 1985

Jastrow, Robert. *God and the Astronomers*. New York: Warner Books, 1908.

Kaufman III, William J. *Black Holes and Warped Spacetime*. New York: W. H. Freedman and Company, 1979.

Kassis, Hanna E. *A Concordance of the Qur'an*. Berkeley, Los Angeles, and London: University of California Press, 1983.

Khathib, M. M. *The Bounteous Koran*. London: Macmillan Press, 1984

Lane, Edward William. *Arabic-English Lexicon*. Beirut, Lebanon: Libraire Du Liban, 1980.

Lovelock, James E. *Gaia, a New Look at the Earth*. London: Oxford University Press, 1974.

Lucas, J. R. *Historical Journal*. XXII (1979): 313-30

March, Robert H. *Physics for Poets*. Chicago: Contemporary Books, 1978.

Marvin, Ursula B. "Continental Drift." *EncyclopaediaBritannica*, Vol. 5:108-115.

Morris, Richard. *Time's Arrows*. New York: Simon and Schuster, Inc., 1986.

Morris, Richard. *The Nature of Reality*. New York: Noonday Press, 1988.

Maududi, Abul A'la S. *The Meaning of the Qur'an*. Lahore, Pakistan: Islamic Publications (Pvt.) Ltd., 1984.

Mayr, Ernst. *One Long Argument: Charles Darwin and the Genesis of Modern Evolutionary Thought*. Cambridge: Harvard
University Press, 1991.

Mayr, Ernst, William Provine, eds. *The Evolutionary Synthesis*. "Prologue: Some Thought on the History of the Evolutionary Synthesis." Cambridge, Massachussetts, and London, England: Harvard University Press, 1981.

Miller, Stanley L., and Robert G. Orgel. *The Origin of Life on the Earth*. Englewood Cliffs: Prentice-Hall, Inc., 1969.

Miller, Kenneth. *Finding Darwin's God*. New York: HarperCollins Publishers Inc., 1999.

Miller, Richard. *The Encyclopedia of Evolution*. New York: Henry Holt and Company, Inc., 1990.

Moyers, Bill. *A World of Ideas*. Edited by Betty Sue Flowers. New York: Doubleday, 1989.

Nasr, Seyyed Hussein. *Introduction, in Islamic Spirituality Foundations*. New York: The Crossroad Publishing Company, 1991.

Nasr, Seyyed Hossein. "Islam and the Environmental Crisis." *MAAS Journal of Islamic Science*. Vol. 6. No.2 (1990): 31-51.

Niazi, Maulana Kauser. *Creation of Man*. Lahore, Pakistan: S. H. Muhammad Ashraf, 1974.

Nicholson, Reynold A. *Rumi*. San Francisco: Divani Shamsi Tabriz, 1973.

Osborn, H. F. *From Greeks to Darwin*. London and New York: The Macmillan Company, 1922.

Pickthall, Mohammed Marmaduke. *Holy Qur'an*. Hyderbad, India: H. E. H. Mir Osman Ali Khan, 1930.

Plott, John C. *The Global History of Philosophy*. Delhi, India: Motilal Banaridass, 1984.

Raup, David M. "Conflict between Darwin and Paleontology." *Field Museum of Natural History Bulletin*, (January 1979): 22-29.

Richard Dawkins. *The Blind Watchmaker*. New York: W.W. Norton, 1986.
River out of Eden. New York: Basic Books, 1995.

Richard Lewontin. A Review of Carl Sagan's book *The Demon- Haunted World: Science as a Cradle in the Dark, New York Review of Books*. (January 9, 1997): 28-32.

Rippin, Andrew and Jan Knappert. *Textual Source for the Study of Islam*. Chicago: The University of Chicago Press, 1990.

Russell, Bertrand. *Religion and Science*. London, New York: Oxford University Press, 1961.

Russell, Robert J. "Theistic Evolution: Does God really Act in Nature?" *Center for Theology and the Natural Science Bulletin*, 15.1, 1995.

Russell, Robert J; William R Stroeger, and Francisco J.Ayala. *Evolution and Molecular Biology.* "Special Providence and Genetic Mutation: A New Defense of Theistic Evolution," by Robert J. Russell.

Vatican City State, Berkeley: Vatican Observatory Foundation and The Center for Theology and Natural Sciences, 1998.

Sagan, Carl. *Cosmos.* New York: Random House, 1980.

Cosmic Connections. New York: Anchor Books, 1973.

"Life." *Encyclopaedia Britannica* 15th ed., Vol. 10:893-911.

Salahi, Adil M., and Ashur A Shamis. *In the Shade of the Qur'an*, Vol. 30. London: M W H London Publisher, 1979.

Schroeder, Gerald L. *The Science of God:The Convergence of Scientific Biblical Wisdom.* New York: The Free Press, a Division of

Simon & Scuster Adult Publishing Group, Copyright © 1977 by Gerald L. Schroeder. Reprinted with permission of The Free Press.

Genesis and Big Bang. New York, Toronto, London, Sydney, Aukland: Bantam Books, 1990.

Shakir, M. H. *Holy Qur'an Translation.* New York: Tahrike Tarsile Qur'an, Inc., 1985.

Shapiro, Robert. "Probing of the Origin of Life." *Encyclopaedia Britannica: Yearbook of Science and Future.* Chicago, 1984.

Shadewald, Robert. *National Center for Science Education Reports.* (March-April 1990) Vol. 10: 21.

Shermer, Michael. "25 Creationists' Arguments & 25 Evolutionists' Answers." *Skeptic.* Vol. 2, No. 2.

Smart, John Jamieson Carswell. "Time," *Encyclopaedia Britannica.* 1981, 15th ed. Vol.18: 410-421.

Stanfield, William D. *The Science of Evolution.* New York: The Macmillan Company, 1977.

Stanley, Steven M. *Macroevolution Pattern and Process.* San Francisco: W. H. Freeman and Company, 1979.

Shreeve, James. "The Argument over a Woman." *Discover.* 2.8 (1990): 52-59.

Strauss, Evelyn. "Can Mitochondrial Clocks Keep Time?" *Science.* (5 March, 1999) Vol. 283: 1436-1438

Trefil, James. *Reading the Mind of Go: In Search of the Principle of Universality.* New York, London, Toronto, Aukland: Doubleday, 1989.

Tufail, Abu Bakr Muhammad. *Hai bin Yaqzan* (The Journey of the Soul). Translated by Riad Kocache. London:U.K: The Octagon Press, 1982.

Turpin R, Lejeune J, Lafourcade J, Chigot PL, Salmon C. (1961) "Presomption de monozygotisme en dépit d'un dimorphisme sexuel: sujet masculin XY et sujet neuter haplo X." *C R Acad Sci[D]* (Paris), Vol. 252:2945-6.

Whinfield, E. H. *Teachings of Rumi.* London: The Octagon Press, 1984. Watch Tower Bible &Tract Society of Pennsylvania (New York). *Did Man Get Here by Evolution or Creation?* 1967y

Yahya, Harun. *The Evolution Deceit.* London, UK: Ta-ha Publishers Ltd., 1999.

GLOSSARY

Adaptation: Physical structure, bodily function (physiology), or behavioral characteristic that strengthens the ability of the organism to cope with its prevailing environment.

Adaptive radiation: Tendency of successful species (or higher groups) to spread into all available ecological niches.

Alga (plural, algae): Any of a broad group of simple, mostly aquatic plants.

Al-Akhirah: The Arabic word for the hereafter. According to Muslim belief our present universe will be completely disposed of and a new universe will be created into which mankind will be resurrected.

Allah: The Arabic term for God, the Supreme Creator of all that exists. There is no plural for the word Allah in Arabic. He has no partners, sons or daughters. He is neither male nor female. He is self-sufficient and no deities precede or exceed him.

Allele: An alternative form of genes occupying the same position (locus) in a homologous chromosome.

Allopatric: Pertains to populations that are geographically separated from one another.

Allopatric speciation: development of new species from single parent species that live in separate territories (allopatric territories).

Amino acid: The basic structural unit of protein.

Amphibian: Any intermediate group of animals between fish and reptiles.

Anagenesis: Transformation of whole populations in the evolution of life.

Analogy (adj., analogous): Similarity in function but difference in structure and origin (e.g., wings of insects and birds).

Annelid: A group of organisms with tiny rings and many body segments fused together in a linear arrangement (earthworms, leeches). They have closed circulatory systems.

Annual ring: Annual variations in the growth rate of trees which can be seen in the series of concentric rings of cross-sections of trees.

Apogee: The most distant point from the center of the earth in the orbit of the moon or of an earth-orbiting artificial satellite.

Arthropod: Segmented animals with paired jointed legs and stiff external skeletons (crabs, lobsters, insects, spiders).

Atom: The smallest part of a chemical element that can take part in any chemical reaction and still retain its identity. All atoms consist of a nucleus and one or more orbiting electrons.

Australopithecus: A genus of fossil hominids found in southern and eastern Africa dating from the Late Pilocene to the end of the Early Pleistocene (at least 2,500,000 years). It is often considered a possible ancestral genus to modern man.

Al-Baari: One of the epithets of God in the Qur'an. An Arabic word derived from the verb *Baara*, which means "evolving from a previously created matter or state." The noun *Al-Baari* means "evolver."

Bacterium: Microscopic unicellular organisms with simple DNA molecules not contained in a nuclear membrane.

Big Bang: The theory that the emergence of our universe was the result of the expansion of matter, space and time, following an explosion of the "cosmic egg" or the initial state of enormous density and pressure at the beginning of the universe.

Big Crunch: Following the Big Bang the universe continues to expand. At some time in the future the universe will start collapsing into a singularity which will lead to its end. This process of corrosion is called the Big Crunch.

Black hole: When a large star implodes due to its own gravitational pull, even the light cannot escape from it. As a result, a black hole is created out of which an observer cannot see the events inside that hole.

Bosons: Subatomic particles we often think of as waves, or radiation. The weak boson carries the weak force when the atomic nuclei decay.

Competition: struggle between members of different species for a mutually required and limited resource.

Chromosome: A long strand of coiled DNA in the nucleus of a cell; threaded around protein. Chromosomes are made up of genes, and each species is classified by its unique number of genes.

Class: A taxonomic grouping of organisms belonging to related orders; the major subdivision of a phylum.

Condensation reaction: A chemical process which takes place when two monomers join. A monomer is a small molecule from which a polymer is derived.

Conifer: A cone-bearing plant.

Coordinates: Mathematical ways of delineating positions.

Cosmic string: When the temperature reaches the so-called Planck value of 1032 degrees (1 followed by 32 zeros), all matter is dissociated into its most primitive constituents, which are called cosmic strings.

Cro-Magnon man: Prehistoric beings considered by some scholars to be the ancestors of modern mankind.

Cytoplasm: The semifluid substance that makes up the nonnuclear part of a cell.

Deoxyribonucleic acid (DNA): The chemical substance in the chromosome the structural arrangement of which is the basis of inheritable characteristics.

Dizygotic: developed from two fertilized eggs (zygotes), as a fraternal twin.

Doppler effect: The displacement of spectral lines (dark lines in a spectrum) in the radiation received from a source due to its relative motion along the line of sight. An approaching motion is indicated by a blue shift; a receding motion is indicated by a red shift.

Dryopithecus: A genus of extinct apelike animals representative of the dryopithecines (a group of small, generalized apes that contains the ancestors of both the modern apes and humans).

Ecology: Study of the interplay of living organisms and their environment.

Echinoderm: A form of marine life with pentaradial symmetry, a calcareous (calcium containing) skeleton, a water vascular system and tubed feet (starfish, sea cucumber, sea urchins).

Ecological isolation: Presence of genetic differences between populations that prevent interbreeding.

Electromagnetic force: The force that operates between particles with electric charge. This is the second strongest of the four fundamental forces.

Electron: A sub-particle with a negative electric charge that orbits the nucleus of an atom.

Ethology: Study of patterns of behavior of animals in their natural environments, with special attention to adaptive and comparative aspects

Event horizon: The periphery of a black hole.

Evolution: Any changes in the frequency of alleles within a gene pool from one generation to another that results in the divergence of related populations and gives rise to new species and higher forms of life.

Fertilization: Union of male and female gametes to form the single cell (zygote) from which the embryo develops.

Fossil: Preserved remains or traces of an organism that lived in the past.

Fundamentalism: A militantly conservative movement in American Protestantism originating around the beginning of the twentieth century in opposition to modernist tendencies. It emphasizes the literal interpretation and the absolute authenticity and accuracy of the Scriptures, the imminent and physical second coming of Jesus Christ, the virgin birth, physical resurrection, and substitutionary atonement.

Fundamentalist: An adherent of Protestant fundamentalism.

Gene: The fundamental unit of heredity with specific sequence nucleotides carried in the chromosomes.

Gene flow: Exchange of genes between local and immigrant populations through interbreeding.

Gene pool: All of the alleles of all the genes carried by individuals in a population at a particular time.

Genetic distance: The extent to which two populations differ consistently in their alleles.

Genetic drift: Change in the genetic composition of a small population resulting from chance event, or from sampling error such as meiosis and random fertilization.

Genetic equilibrium: Invariability of particular frequencies of allelic members of a gene through successive generations in a population.

Genetic isolate: A self-containing breeding population that does not exchange genes with any other group.

Genotype: The full set of genes of an organism's chromosome, as distinct from his or her outward appearance (phenotype).

Genus (plural, genera): A taxonomic grouping of very similar organisms considered to be closely related species.

Geographic barrier: A geographical feature (such as mountain or river) that prevents gene flow between populations.

Gluon: A subatomic particle of zero mass. Quarks interact through the exchange of gluons that carry the strong nuclear force.

Grand Unified Theory: A theory that aims to unify the basic forces of nature, bringing into one scheme the strong and weak forces, electromagnetism and gravity. At the extraordinarily high temperatures prevailing in the earliest moments of the Big Bang, these forces were indistinguishable from each other.

Gravitation: The force that keeps the entire universe together by mutual attraction.

Graviton: A massless subatomic particle. The graviton carries the force of gravitation.

Hadiths: The reports of what Prophet said and practiced.

Hemoglobin: Iron-containing protein in vertebrate blood that binds and transports respiratory gases (oxygen and carbon dioxide).

Heredity: Genetic transmission of traits from parent to offspring.

Hominid: Member of the family of upright, bipedal primates that includes modern humans and their related ancestors.

Homo erectus: The extinct species of hominids (the family of man), dating from the middle Pleistocene time (2,500,000 to 10,000 years ago).

Homology: Similarity of structures in a series of related organisms because of descent from common ancestors, without regard to function.

Homologous structure: Body parts that are similar in structure owing to common evolutionary and developmental origins.

Homo sapiens: The species to which all modern human beings belong.

Homo sapien sapiens: Modern human.

Iblis: The Arabic proper name for the Devil.

Immunity: The ability of an organism to resist or overcome an infectious agent or its antigens by the presence or production of antibodies.

Inbreeding: The mating of related individuals with similarities in their genetic makeup. This mating occurs in greater or lesser frequency than would be predicted by chance.

In vitro: Within a glass; observable in culture or in a test tube.

Isolating mechanisms: A morphological, behavioral, physiological, or geographical barrier that prevents or limits gene exchange between different species.

Jennat-ul-Khuld: The Arabic term for Eternal Paradise; the place of reward and final return for those who are righteous in their worldly thoughts and actions.

Kelvin scale of temperature: A scale of temperature whose units (called kelvin, symbol K) are equal in size to those of the Celsius scale, whose zero is fixed at -273.16°C, often known as absolute zero.

Khalaqa: A verb used in almost all verses of the Qur'an in that refer to the creation of mankind. It means "to bring into existence according to a certain measure or proportion a being, and to make it equal to another thing which is not like any preexisting being."

Linkage: An association of genes of different characteristics in the same chromosome. In the absence of crossing over, linked genes do not assort randomly.

Mammals: The highest class of vertebrate animals that possess hair and suckle their young.

Masnavi: A collection of the spiritual couplets of Jalaluddin Rumi, which has greatly influenced Muslim mystical thoughts and literature.

Meiosis: The cellular process that results in the number of chromosomes in the gamete-producing cell being reduced to one half (the process of the formation of sperm and egg).

Melanin: Dark brown or black pigment of the skin or hair.

Micro-organism: Minute, usually microscopic, living organisms such as bacteria, viruses, and yeast.

Milky Way: The name of a great arc of stars and dust, seen in clear skies from most latitudes, in or near the plane of the galaxy in which the Sun, Earth, and other members of the solar system are located.

Monomer: small molecules from which a polymer is derived.

Mono-zygote: Developed from a single fertilized egg (zygote) as identical twin.

Musawwir: An Arabic epithet of God used in the Qur'an. It means, "Fashioner, of all existing things. The One who created them and gave each of them a special form and specific manner of being whereby the can be distinguished from others." In modern biological vernacular, it could mean "someone who causes speciation."

Mutation: An alteration of the gene, by deletion, addition, or substitution of one or more nucleotide bases of DNA strand, which results in an inheritable change in its traits.

Natural selection: Differential reproduction of offspring (genotype) from one generation to the next.

Neanderthal: A type of prehistoric creature who inhabited much of Europe and the areas surrounding the Mediterranean during the earlier Late Pleistocene Epoch. The Neanderthals were cave dwellers, short but powerfully built.

Niche: The way of life of a species that includes every aspect of its ecological, functional, and behavioral roles.

Nuclear decay or radioactivity: The spontaneous breakdown of an unstable atomic nucleus with the emission of particles and rays.

Nucleic acid: long-chain polymer of nucleotides.

Nucleotide: A basic repeating unit in a nucleic acid consisting of a 5-carbon sugar molecule, a phosphoric acid, and a ring-shaped nitrogenous base.

Nucleus: In atomic physics the nucleus is the central part of an atom, consisting only of protons and neutrons, held together by a strong force. In biology, the nucleus is the spheroid body within the cell that is bound by a double membrane and contains the chromosomes.

Neutron: An uncharged particle in the nucleus.

Neutrino: A subatomic particle with no mass and no electric charge; it enters only into reactions involving the weak nuclear force.

Orbit: The path of one body around another.

Order: A taxonomic grouping or organisms belonging to similar families.

Organic compound: A chemical compound that has one or more carbon atoms.

Oxidation: The removal of electrons from an atom or compound; oxidation in biological systems generally involves removal of hydrogen electrons.

Paleontology: The study of the past through its fossil remains. A practitioner of the science of paleontology is called paleontologist.

Phalanx: A bone of a finger or toe.

Phenotype: outward (observable) manifestation of individual genetic makeup.

Photon: A quantum particle of light and the messenger particle of electromagnetic radiation (a particle of visible light).

Phyletic gradualism: The notion that evolutionary change is slow and steady.

Phylum: The broadest grouping below kingdom, characterized by a structural plan that differs fundamentally from and is difficult to derive from any other.

Polymer: Large molecule made up of many repeating or like subunits called monomers.

Population: Individuals of a species that form a local breeding community in a particular region.

Proton: The positively charged particles that make up roughly half the particles in the nucleus of most atoms.

Protozoa: Any of the single-celled organisms.

Punctuated equilibrium: The concept that species (and higher groups) remain stable for long periods of time (equilibrium), then change rapidly (punctuation).

Quantum: The indivisible unit in which waves may be emitted or absorbed.

Quark: A charged elementary particle that feels the strong force.

Protons and neutrons are each composed of three quarks.:

Qur'an: Islam's book of spiritual guidance. Muslims believe that the Qur'an, in its original Arabic vernacular, is the word of God, revealed to the Prophet Muhammed over a period of twenty years. Translations and commentaries of the Qur'an in any language reflect the translators' interpretations and should not be confused with the Holy Book.

Rabb: An Arabic epithet of God used in the Qur'an. Its literal meaning is "someone who fosters a thing in such a manner as to make it attain one condition after another until it reaches its goal of perfection."

Radioactivity: The spontaneous breakdown of the nuclei of isotopes into the nuclei of other atoms, accompanied by the emission of energy in the form of atomic particles. This breakdown or decay process of nucleus is random, for each kind of radioactive nucleus there is a specific time period, called half time, which describes its rate of decay. In a large sample of nuclei, almost exactly half will have decayed in one half life.

Radioactive dating: The radioactive elements provide accurate nuclear clocks; by comparing the relative abundances of a remaining radioactive element and the element it decays to, we can learn how long the process has been going on and hence arrive at the age of the sample.

Rakah: A fixed series of ritual movements and recitation that are repeated in the daily mandatory prayers of Muslims.

Ramapithecus: This is a fossil primate genus dating from the late Miocene or early Pliocene Epoch classified as a member of the family Hominidae, which includes Australopithecus and Homo, the genus to which modern mankind belongs.

Random mating: Pattern of breeding in which an individual of one genetic makeup has an equal probability of mating with any individual of another genetic constitution.

Rinonucleic acid: (RNA, Nucleic acid based upon ribose; occurs in three forms: messenger RNA (mRNA), Ribosome RNA (rRNA), and transfer RNA (tRNA).

Red shift: The reddening of light from a star that is moving away from us, due to the Doppler effect.

Reducing atmosphere: Atmosphere without oxygen.

Reduction: The gain of an atom or molecule by an electron in a chemical reaction.

Regulatory gene: A gene that controls the expression of one or more other genes.

Reproductive isolation: Inherent inability of members of different breeding populations (particularly different species) to breed with each other.

Ribosome: A small organelle in the cell. It is the site of protein synthesis.

Scientific method: The organized approach to the study of natural phenomenon. The initial step in the process is observation and it is followed by the development of hypothesis, systematic experimentation to test the hypothesis, and finally the formulation of an explanation for natural phenomenon based on the results of the experiment.

Sedimentary rock: Rock formed of sediments (particles) of chemical or organic origin.

Singularity: Where space-time is reduced to a mathematical point and the density of matter becomes infinite.

Species: A group of similar organisms whose members can breed with one another to produce fertile offspring.

Stratigraphy: The study of layers of rocks, particularly with respect to the relative ages of the fossils they contain.

Strong force: The strongest of the four fundamental forces with the shortest range of all. It holds quarks together within protons and neutrons and holds protons and neutrons together to form atoms.

Subspecies: A geographical race of a species.

Taxon (plural, taxa): A group of organisms such as species or class as defined by the biological classification scheme (taxonomy).

Taxonomy: Study of classification of organisms into appropriate categories (taxa) on the basis of evolutionary relationships among them. The system consists of a hierarchy of categories. The most inclusive categories are the kingdoms. Other main categories, in descending order, are phylum, class, order, family, genus, and species.

Theory: Hypothesis that is supported by extensive observations and experiments.

Time dilation: The time between two events (the sending of light and its reception at the receiver) is greater for an earth-bound observer than for a traveling observer. This is a general result of the theory of relativity. Its effect is that time is actually measured to pass more slowly by the clock of a traveler (astronaut) than by the clock of a stationary person (earth-bound person).

Uniformitarianism: A theory that that finds all features of the present earth can be explained in terms of the accumulated effects of processes that we can see and measure here and now.

Vertebrate: Animal possessing a backbone.

Vestigial organs: Refers to an apparently functionless or imperfectly developed structure or organ that was fully developed and functional in ancestral species.

Weak force: One of the four fundamental forces experienced by elementary particles except the force-carrying particles. It is responsible for radioactivity.

Xylem: Complex vascular tissue that conducts water and minerals from the roots to other parts of plants.

Zygote: Diploid cell resulting from the union of two haploid gametes (fertilized eggs).

CPSIA information can be obtained at www.ICGtesting.com
Printed in the USA
LVOW052052210612

287068LV00002B/207/P